The Art of Strategic Execution is the best playbook I've read on the underrated yet indispensable role of TPM in technology companies. Priyanka Shinde offers an optimal combination of practical tips, real-life stories and research–a real force multiplier for anyone interested in launching successful software and hardware products.

– BILL DEMAS,
Entrepreneur, ex-CEO of Conviva, Shopkick, Turn

In *The Art of Strategic Execution*, Priyanka Shinde distills two decades of wisdom gained in fast moving high tech companies. It's not just a book; it's a passport to the world of Technical Program Management!

– ALI DASDAN, **PhD**,
EVP & CTO of Zoominfo

Given the industry's confusion on what a TPM actually does, the book is clarifying about both what a TPM is and is not. Shinde also addresses one of my personal favorite topics, how a high functioning TPM can accelerate engineering productivity and quality, building a stronger engineering culture. I will encourage everyone who joins my team to read and reference it.

– SOPHIA VICENT,
Senior Director of Engineering Operations, Doordash

The Art of Strategic Execution is a wonderful combination of career advice, practical knowledge, and useful frameworks for execution and communication... a great resource for TPMs as well as anyone working across complex programs.

– **MANLIO LO CONTE,**
Chief Product Officer of Zwift

Finally, the most comprehensive book for TPMs is here! *The Art of Strategic Execution* is a reference guide that you can keep returning to—ideal for applying skills to specific situations or leveraging frameworks. Imagine having a coffee chat with Priyanka Shinde, drawing insights from her experience at tech companies known for exceptional execution, such as Meta. A solid companion for every TPM.

– **SANDEEP KACHRU,**
Head of Data Platform, Airtable, ex-Meta, ex-Google

The Art of Strategic Execution is insightful, approachable, and lucidly written. It is full of practical strategies and actionable advice grounded in real-world experiences. This book pushes me to be more intentional in my mentoring tasks and I recommend it for early career engineering academics who may lack formal leadership and management training.

– **ARNAV JHALA, PhD,**
Associate Professor of Computer Science,
North Carolina State University

This book has no fluff, just straight talk! Priyanka Shinde's personal journey forms a compelling narrative, like getting 1:1 mentorship for managing your career. From acing interviews to leveling up through meaningful impact and achieving lasting success, this book exudes pragmatic wisdom for immediate use. A must-read for all TPMs and tech leaders!

– ASHWIN RAMACHANDRAN,
Head of Engineering, Interview Kickstart

Priyanka Shinde provides a compelling treatise on the role of a TPM and how they fit into, and drive, today's technology teams. *The Art of Strategic Execution* speaks to the novice TPM, seasoned professionals, and mature teams to guide and evolve their understanding of the TPM role.

– SIDDHARTH JOSHI,
Senior Manager, Program Management, Expedia Group

The Art of

_of_

STRATEGIC EXECUTION

Finding Success in
Technical Program Management

PRIYANKA SHINDE

This book may be purchased in bulk for education, business, fundraising or sales promotional use. For information, please email booksales@artofstrategicexecution.com.

ISBN 979-8-9896722-0-2 (hardcover) | ISBN 979-8-9896722-1-9 (paperback)
ISBN 979-8-9896722-2-6 (Ebook)

Library of Congress Cataloging-in-Publication Data

Names: Priyanka Shinde, author.

Title: The Art of Strategic Execution : Finding Success in Technical Program Management / Priyanka Shinde.

Description: San Francisco, CA : Includes bibliographical references

Identifiers: LCCN 2024900326 | ISBN 979-8-9896722-0-2 (hardcover) |
ISBN 979-8-9896722-1-9 (paperback) | ISBN 979-8-9896722-2-6 (Ebook)

Subjects: LCSH: Shinde, Priyanka | Project Management | Leadership |
Career development | Management |

The web addresses referenced in this book were live and correct at the time of the book's publication but may be subject to change.

Printed in the United States of America

This book reflects the author's personal opinions, and does not represent the views of her past, present or future employers.

This work depicts actual events in the life of the author as truthfully as recollection permits. While all persons within are actual individuals, names and identifying characteristics have been changed to respect their privacy.

To Sandeep and Neev
For always championing me

To the TPM Community
You got this!

This book, *The Art of Strategic Execution* is a step towards a monumental aspiration of mine where there will come a day when everyone understands the value of Technical Program Management and no one needs to ask, *"What is a TPM and what do they do?"*

TABLE OF CONTENTS

INTRODUCTION

Picture a 14-year-old introverted girl who has just moved to a new city. Going to a big city high school is a significant change for someone who, for the past eight years, had been in familiar surroundings, with the same classmates, in the same school, and in the cozy comfort zone. I was that quiet, studious kid who always got good grades and hung out with just a few close friends. I didn't seek the limelight, never asked too many questions, and rarely ventured into extracurricular activities because, in my mind, those weren't for "serious" students.

Back then, I used to take the bus to school with the same group of kids. One day, one of the younger girls at my stop casually said, "You're very *timid*." I didn't think much of it at the time, but that word stayed with me as the years went by. Recently, during a reunion with a colleague from two decades ago, he recalled, "You were so quiet and timid back then." Once again, that label echoed in my life. That was my reality.

My journey into the world of computer science and software engineering suited my introverted nature. It allowed me to excel without needing to interact with people much. I took pride in doing my work well, but I never sought recognition or advocated for myself. In my stint as a software engineer at a company on the east coast of the United States, I got a chance to work with program managers. They were responsible for managing the software launch, coordinating with various teams, and making things happen, and I found their role fascinating. I believed my technical background and natural organization skills could make me a great fit for that role. What impressed me even more was how well-respected program managers were in my team, which led me to believe everyone understood their value and role. I saw the potential for myself and decided that becoming a Technical

Program Manager (TPM) would be a fulfilling career path. Like any problem I'd tackled before, I approached this pivot with thorough research and a hunger for knowledge. I believed that adding an MBA to my education, which included a Masters in Computer Science and a Bachelors in Computer Engineering, might boost my credibility a bit. Armed with both master's degrees, I set my sights on opportunities in the tech hub of Silicon Valley, the epicenter of the industry. Back then, the role of a TPM wasn't as clearly defined as it is today and the tough job market didn't help either. Finally, after countless interviews and a year and a half of relentless effort, I officially became a TPM.

Initially, I felt confident in my abilities to collaborate with technical teams, manage programs, and deliver results. However, my introverted nature, which had served me well as an engineer, didn't quite align with the demands of technical program management. The culture in Silicon Valley required me to step out of my shell and be more assertive. When I started at a large company, the consequences of not doing so were significant and I realized I needed to shift my mindset. My success hinged on my ability to get out of my comfort zone and stretch myself. Today's modern tech companies give a lot of autonomy to individual teams, making the TPM role quite challenging. It's no longer about just following top-down mandates; you need to lead, influence, and cultivate relationships. Through my TPM career journey, from startups to renowned companies like Meta (formerly known as Facebook) and Cruise, I kept working on my role as a TPM. I deepened my technical skills, polished up my program management expertise, and worked on being a strong leader. I started to proactively advocate for myself, my role, and communicate my technical and business ideas effectively to gain support and recognition. My hard work paid off, achieving results and illustrating the indispensable value of soft skills in the success of TPMs.

Through my journey of more than a decade in technical program management, I have consistently observed how many

TPMs get bogged down in the mechanical aspects of the role—like project management tools, Gantt charts and certifications—and neglect developing their leadership and communication skills that set apart exceptional TPMs from average ones.

My experience has set me up to understand deeply what it takes to become a successful TPM, and how to guide others to succeed while making a meaningful impact on the organization. I have had the privilege of launching large-scale cutting edge products and building TPM teams from the ground up. Each time I joined a new team, I took on the responsibility of bootstrapping the TPM function, defining and evangelizing the role, and working with the TPM community to establish the TPM brand. Mentoring, coaching, interviewing, recruiting, and connecting with other TPMs became an integral part of my journey. At Facebook, I initiated several programs such as the Aspiring TPMs Circle and TPM Mentorship, and streamlined TPM interviewing and recruiting practices within the Assistant and Reality Labs organization. At Cruise, I had the opportunity to build a world-class TPM organization to deliver some of the most complex products. I developed new engagement models to bring clarity and situated TPMs to own programs and set them up for success by providing growth opportunities. My passion for the TPM function and the value it brings to teams and organizations pushes me forward today. I firmly believe that when leveraged correctly, TPMs can have a profound impact on their team's success.

My journey, from that timid girl on the bus to a successful TPM leader in Silicon Valley, holds lessons I'm eager to share. In *The Art of Strategic Execution*, I'll dive into the intricate world of TPM and share the insights and wisdom I've gained along the way. I will provide you with tips, tricks, and actionable strategies. I am also excited to bring you stories and experiences from other successful TPMs and TPM leaders; as well as business leaders who are staunch supporters of TPM.

I want to emphasize what this book is not. It does not aim to teach system design or technical concepts, nor does it focus on program management methodologies like Agile or step-by-step instructions for performing actions like estimation, timeline development, or setting up program management tools. Instead, this book emphasizes the art rather than the science of program management and execution in relation to the demands of modern tech companies. It explores the intangible aspects—the mindset, communication, influence, and leadership—that set exceptional TPMs apart. You will also find various templates throughout the book that will help you build a strategic thinking mindset and streamline your execution. This book is a guide to finding success for anyone navigating the complex world of technical program management with confidence, adaptability, and excellence.

I also encourage organizations and cross-functional leaders who work with TPMs to utilize this book to create strong partnerships with TPMs and empower them for overall business success.

Join me as we explore *The Art of Strategic Execution!* Together, we will unlock a successful career in the world of Technical Program Management.

Part 1

INTRODUCTION TO TECHNICAL PROGRAM MANAGEMENT

*If I could collect a penny every time someone asks
"What is a Technical Program Manager?",
I would be a millionaire!*

CHAPTER 1

WHAT IS A TECHNICAL PROGRAM MANAGER?

TPMs are chaos-destroying machines,
and each new person you bring onto your team,
each dependency you create,
adds hard-to-measure entropy to your team.
A good program manager thrives on measuring,
controlling, and crushing entropy.

– Michael Lopp

I vividly recall the project that served as a pivotal moment in my career. At the time, I was working as a dedicated software engineer writing code for a student and financial information software. I was part of a big team of 20 engineers. The company was experimenting with new technology and management was looking to build a small web service application as an add-on feature for customers. My manager knew I was learning this new technology on the side and so this exciting project landed in my lap. I was asked to lead development and had one other engineer assigned to the project. Even though the software coding aspect was not immense, it required me to direct the business aspects of the product launch because there were no other dedicated resources. The project started with just one line problem statement that was not well-defined. This gave me the opportunity to get involved in the concept stage - defining requirements, identifying success criteria, and even designing the user interface. In addition to coding the web service with my fellow engineer, I had to troubleshoot issues with integrating the service with our external vendor. Since this was a brand new technology, I had to devise a meticulous deployment plan in collaboration with the IT department. Finally, I had to formulate an effective go-to-market strategy hand in hand with the marketing team.

Being involved in aspects beyond coding and collaborating with multiple teams across the organization gave me insight into different types of complexity that exist in the product development lifecycle. The ability to have full ownership of the problem space instilled in me a deep sense of fulfillment. The sheer exhilaration I experienced while working on this project gave me a taste of what it meant to manage a technical program end-to-end. At that time, my observations of the value brought on by program managers had already piqued my interest in the role of a TPM. However, it was the hands-on experience of launching a new web product that truly solidified my decision to shift my career path and enter the world of technical program management. Making the pivot wasn't easy as there weren't a lot of resources and TPM wasn't a common job title. So I started looking for different program manager types of opportunities that mentioned working with engineering teams. I built relevant experience in my current job, but it took me at least another four years to make the official transition.

It has been over a decade since that web service project and my passion for technical program management remains unwavering. It was during my initial days at this one startup where the questions came up about what a TPM does and how it differs from other roles. That's when I thought about researching the title so I could present a cohesive answer and write down clear roles and responsibilities. Since then, I've frequently explained the TPM role and its duties to new team members. I have had the opportunity to be the first TPM on a team multiple times, grow TPM teams, and set standards for the function. These experiences have brought me to this point, where I can share insights into what it entails to be a TPM and how to achieve success in Technical Program Management.

EVOLUTION OF THE TPM ROLE

My first TPM role was actually advertised as Program Manager, but it was highly technical in nature. The Program Manager role was common at Microsoft, but it wasn't called TPM, nor did I come across any formal definition or distinction. I always thought "Technical" was added to distinguish the area of focus. The terms TPM, Project Manager and Program Manager are often used interchangeably, but there are nuances that make the roles unique from each other.

PROJECT MANAGER (ALSO REFERRED TO AS PjM)

Manages a single project and the teams responsible for fulfilling that project at a given time. **A project can be defined as an endeavor with a finite start and finish, with a primary focus on execution.** The goal is to launch deliverables with resources available within schedule and budget. A PjM's role is more tactical, with a focus on operational elements such as meeting deadlines, staying within budget, and delivering on time. Their work is complete when the project is complete.

Example: Collecting local business data as part of a Google Maps Local product for a new city.

Both Project and Program Managers are usually brought in once the requirements are finalized. While they require relevant domain knowledge, their core focus is on getting tasks done on both time and budget.

PROGRAM MANAGER (ALSO REFERRED TO AS PgM)

Manages **a program, which is a collection of multiple interdependent projects that need to be coordinated together.** The scope is usually larger, includes ongoing iterative changes, and can last longer across multiple phases. PgMs create program level plans to achieve a combined goal of one or more programs. They collaborate with multiple teams and other project managers to understand status across different projects and support action to improve delivery. The PgM is usually seen as a leader for the entire program. They can articulate goals and objectives and their impact on the business. They may also look into long term program goals and continue to be involved in future phases. The PgM role evolved from PjM to reflect the extended scope and potentially different skill set needs.

Example: Launching Google Maps in a new country requires multiple projects across different teams and even working with external vendors. Depending on the number of features being launched in the initial phase, this program can span multiple years as more services may be added over time.

The roots of the TPM role date back to the 1980s when the software industry started using project management principles. "Project management has been practiced for thousands of years since the Egyptian era, however, it has been about half a century ago that organizations start applying systematic project management tools and techniques to complex projects. In the 1950s, the Navy employed modern project management methodologies in their Polaris project. During the 1960s and 1970s, Department of Defense, NASA, and large engineering and construction companies utilized project management principles and tools to manage large budget, schedule-driven projects. In the 1980s, manufacturing and software development sectors started to adopt and implement sophisticated project management practices. By the 1990s, the project management theories, tools, and techniques were widely received by different industries and organizations" (Carayannis, Kwak, and Anbari 2005, 1-9).

Over the next few years, it became apparent that software was not like other industries and its complexities required the project or program manager to have a technical understanding of the systems and architecture. Thus came about the birth of the TPM. It is essentially an evolution of the Program Manager role where the primary focus is on managing technical programs and working closely with engineering teams. The incoming TPM was required to possess a deep technical domain knowledge so they could have productive conversations with engineering and effectively manage technical programs. Many of the early TPMs came from engineering backgrounds and typically had expertise in specific technology, which they leveraged to design, create, and deliver business objectives.

TPMs often manage programs that vary from deeply technical to complex, multi-year, strategic initiatives. These programs may span multiple teams, organizations, or even companies. In order to be effective, TPMs required different skills than program managers. TPMs are meant to be problem solvers with excellent leadership skills who can motivate the team to reach all business objectives. They need to understand team dynamics, product ecosystem, and technical landscape. They often work on products/features that are technically complex and require extensive cross functional collaboration.

Here's an example of a program from my TPM career that clearly demonstrates the important role TPMs play in managing large, complex, and technical programs. This program aimed to launch the first-ever conversational Artificial Intelligence (AI) product on a brand-new hardware device being developed by the company - a first for the team. I worked with 50+ engineers across three different organizations. Less than six months from the public release date, the hardware team was still resolving design updates, the speech team was still improving model accuracy, and my conversational AI team was still figuring out how to match the user experience with the model metrics. We were running into quality issues like response lag and timeouts. Since this was the first time we were integrating existing social connections into the AI product, we kept discovering issues that degraded user experience. With so much ambiguity, a tight deadline and technical complexity, this program was perfect for a TPM to reign in the chaos. In fact, it could not have been successfully launched within the given deadline without a TPM.

WHAT DO TPMS DO?

In reality, no universal definition of the TPM role exists. Every company that has hired TPMs has created its own version of the role and every team has morphed it slightly, or even mangled it. Having worked in various places big and small, I find that providing a mission statement works better to help TPMs and their partner teams understand what they do and why.

> **Technical Program Managers** drive **complex** and **cross-functional technical** programs by leveraging **deep domain expertise**, building a **holistic execution strategy**, and harnessing **human leadership** qualities to deliver **strategic business outcomes** that have a wide-reaching impact.

TPM Mission Statement

Amazon was probably one of the first companies to use the Technical Program Manager title and put together a well-defined TPM structure. The Microsoft Program Manager role was essentially a mix of product and program management, which has now been split formally into Product Manager and TPM. The TPM title has also spread to other industries such as banking, healthcare, and consulting.

Since Amazon's success with TPMs, many more companies of varying sizes have adopted the TPM role, which led to its prominence today. In the last decade, increasing technological advancements continue to shape and grow the role. The products and services built now are more complex than a couple of decades ago, and such complexity requires an in-depth understanding of the systems and architecture. The role has evolved from tactical aspects involving coordination, facilitation, and task creation to strategic execution requiring technical expertise, domain knowledge, cross-functional alignment, and risk management.

* * *

PRO TIP: When seeking new opportunities as a TPM, don't get tied to the title. Instead, focus on the responsibilities and how they align with the core concept of TPM.

CHAPTER 2

TPM VS. OTHER TECH ROLES

Clarity creates an organizational road map to success. It drives faster and better decision-making while increasing trust.

– Tim Leman

The running joke within some engineering teams is how many different ways can you arrange the letters T, P, and M to create new role titles. It can get confusing not just for engineers, but even for TPMs. Some organizations started using the acronym TPgM to indicate a function that is a mix of TPMs and PgMs. The "g" in PgM itself was added so there wasn't confusion with the PM for Product Manager.

Here are the different roles that often appear in Tech. All these roles intersect and overlap with each other in various ways. They also play a complementary role to each other depending on the team and program circumstances.

- Technical Program Manager (TPM)
- Product Manager (PM)
- Product Manager Technical (PMT)
- Engineering Manager (EM)
- Program Manager (PgM)

- Project Manager (PjM)
- Technical Project Manager (TPjM)

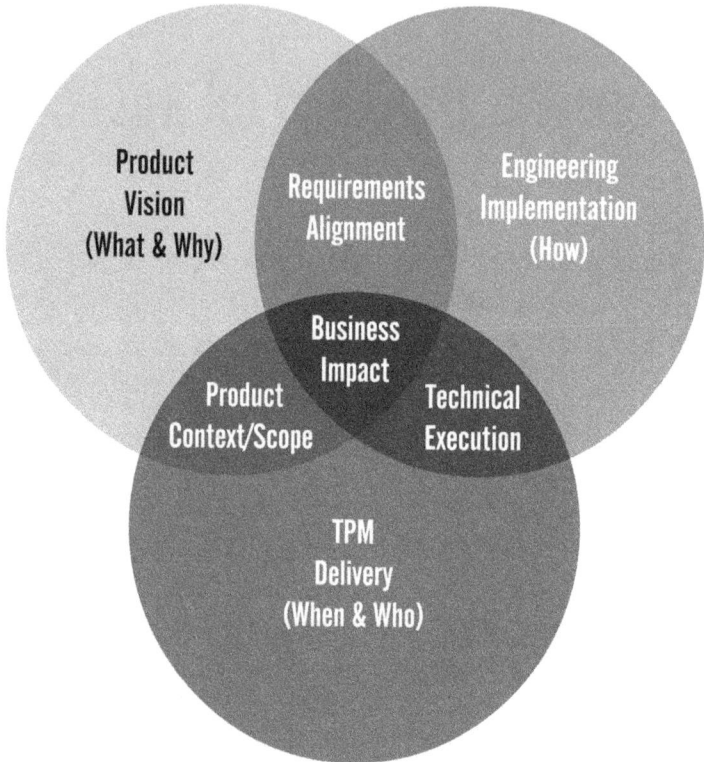

Product
Vision
(What & Why)

Requirements
Alignment

Engineering
Implementation
(How)

Business
Impact

Product
Context/Scope

Technical
Execution

TPM
Delivery
(When & Who)

TPM-PM-Engineering Intersection View

TPM VS. PRODUCT MANAGER (PM)

In companies that specialize in building high-tech products and services, the roles of Product Manager and Technical Program Manager often coexist. There are many overlapping skills and responsibilities between these two roles, which can cause confusion and conflict. I have worked with different types of PMs and there isn't a magic line that can be drawn to distinguish these two roles as they frequently intersect and complement each other.

In one such conflict situation, I worked on creating a highly detailed RACI (Responsible, Accountable, Consulted, and Informed) table. I utilized the simplest way to differentiate between the two - which is to say that PMs define the "what" and the "why" while TPMs define the "when" and, to some extent, the "how." However, that wasn't enough because there is more to creating a collaborative working relationship than a RACI chart. If both the Product and Program Manager want to do the same things, creating a detailed breakdown may not help.

That's why I prefer the following articulation:

"PMs lean more towards vision and strategy,
while TPMs lean more towards execution and delivery."

For programs where both TPMs and PMs are involved, this articulation is an attempt to help the teams and people in these roles understand their uniqueness and similarities.

The emphasis on the word "*lean*" implies that each role has a core value proposition, but can also encompass aspects of the other. PMs should have knowledge of execution and possess some technical domain expertise, while TPMs should understand the vision and strategy and possess effective prioritization skills. You can view it as a sliding scale or a see-saw. You may lean into one area more, depending on the program requirements and the team composition. In complex programs, the partnership between TPMs and PMs can be especially valuable. While PMs are forward-looking, envisioning ways to bring their vision to life while considering what lies ahead, TPMs are laser-focused on present execution and ensuring that tasks are completed. This complementary dynamic allows for a balanced approach, addressing both the strategic and tactical aspects of program management.

Here are four ways that PMs and Technical Program Managers can collaborate and complement each other:

- **Shared vision:** PMs and TPMs align on the overall product vision and strategic objectives, ensuring that the program execution supports the product strategy.

- **Requirements and execution:** PMs provide detailed product requirements and priorities, while TPMs bring their technical expertise to drive the execution and delivery of those requirements.

- **Cross-functional collaboration:** PMs and TPMs work together to facilitate seamless collaboration across teams, ensuring that the product vision is effectively translated into actionable plans and successful execution.

- **Feedback loop:** PMs gather user feedback, market insights, and data on product performance, providing valuable input to TPMs for refining and improving program execution strategies.

In my experience, to establish a fruitful partnership with your TPM or PM, it is crucial to engage in open conversations and clearly outline your respective roles and responsibilities, taking into account your individual strengths. By openly discussing expectations, aligning on objectives, and leveraging each other's expertise, you can foster a collaborative environment that maximizes the potential of both roles.

The best partnership I had with a Product Manager was with Lira. We both jumped into a program that was already underway, highly chaotic, and behind schedule. We connected with each other to understand the status of the program and what was needed. Together, Lira and I discussed our roles and responsibilities, and what each of us brought to the table that was complimentary to the other person's strengths. We leveraged each other every day and asked for help if something urgent was coming up. As a team, we built a great deal of trust, so we never had to worry about stepping on each other's toes.

Which Role Is Right for Me?

Many folks often move between the TPM and PM role until they figure out which one is for them. If you are thinking about which of these two roles is right for you, then it is important to understand that PMs and TPMs possess distinct characteristics, responsibilities, and skill sets that contribute to their effectiveness in driving successful outcomes.

Product Manager (PM)	Technical Program Manager (TPM)
• **Vision:** PMs have a forward-thinking mindset, envisioning the future direction and growth of the product or service. • **Strategic Thinking:** They excel in formulating and executing strategic plans aligned with market trends, customer needs, and business objectives. • **Customer-Centric:** PMs thoroughly understand user needs, conduct market research, and leverage insights to deliver products that resonate with customers. • **Messaging:** PMs focus on positioning their product in the market effectively and crafting compelling product messaging. They also implement strategies to monetize the product, and optimize revenue generation.	• **Breaking Down to Details:** TPMs pay meticulous attention to program details, ensuring that execution aligns with technical requirements and specifications. • **Execution Tactics:** They have a relentless focus on completing tasks, managing schedules, and ensuring successful program delivery within the defined timelines. • **Cross-Functional:** They can collaborate and lead cross-functional teams, fostering collaboration and alignment across various stakeholders. • **Smooth Delivery:** TPMs excel in analyzing complex challenges, identifying potential risks, and formulating effective mitigation strategies.

PM vs. TPM Focus Areas

PMs often immerse themselves deeply in the product or area they are working on. They identify significant opportunities, conduct research and analysis, define priorities, and craft detailed requirements. On the other hand, TPMs can have a narrower or broader focus depending on the domain area. Regardless of the specific focus, both roles demand strong communication and leadership skills to effectively collaborate with stakeholders (other teams, business groups, etc.) and drive successful outcomes. While they have unique focuses, they can collaborate and complement each other in various ways to achieve shared goals.

Product Manager (PM)	Technical Program Manager (TPM)
• **Product Strategy Definition:** PMs shape the overall product vision, set strategic objectives, and prioritize features to deliver maximum value to customers. • **Roadmap Planning:** They create product roadmaps, establish release schedules, and prioritize features based on business objectives and customer needs. • **Market Analysis:** They conduct market research, identify customer pain points, analyze competition, and explore emerging trends to inform product decisions. • **Requirement Gathering:** PMs work closely with stakeholders, including customers, engineers, designers, and marketers to gather and document product requirements.	• **Program Planning:** TPMs collaborate with stakeholders to define program goals, scope, and deliverables, ensuring alignment with strategic objectives. • **Execution Oversight:** They coordinate program activities, manage timelines, track progress, and proactively identify and address bottlenecks or risks. • **Risk Management:** TPMs identify potential risks and develop mitigation strategies, working closely with technical teams to ensure robust contingency plans are in place. • **Communication and Reporting:** They facilitate effective communication among team members, stakeholders, and executives, providing regular updates on program status, milestones, and key metrics.

PM vs. TPM Responsibilities

If your current job/situation allows, I encourage you to experiment with both roles in order to land on something that compliments your strengths. Remember that you may be going back and forth between the nuances of these roles within the same program. Some programs or teams may not need both roles and in that case, you will wear both hats. The key is to be flexible.

TPM VS. TECHNICAL PRODUCT MANAGER (PMT)

The Technical Product Manager, also known as Product Manager Technical or PMT for short, has emerged in the last 5-7 years. I found myself right in the middle of this evolution once more, experiencing it firsthand at multiple companies. When I joined a big startup team, I learned that both Technical Program Manager and Technical Product Manager existed with the same abbreviation - TPM. To add to the confusion, they used to be the same team, but separated out at some point. Nevertheless, engineering did not understand the differences or the reasons behind the split. I took on the task of bringing a sense of clarity to these roles. This resulted in the PMT title being adopted officially, along with a clarification on the need for both roles in the context of the company's initiatives.

The rise of the PMT role can be attributed to the increasing complexity of products and systems in the tech industry. The PMT role is similar to a traditional PM role, but with an added requirement of having a technical sense to work with complex systems. Organizations working on advanced technologies like AI, Machine Learning (ML), and autonomous systems require individuals who can bridge the gap between technical intricacies and business objectives. PMTs ensure that products are not only innovative but also feasible, and aligned with technical constraints. Organizations have recognized the need for specialists who possess both product management acumen and technical prowess. The PMT role emerged as a response to this demand. Having individuals dedicated to understanding the technical underpinnings of products allows organizations to make informed decisions, reduce friction between technical and non-technical teams, and streamline the development process.

Furthermore, the distinction between TPMs and PMTs in terms of titles helps organizations hire more strategically. It allows hiring managers to clearly communicate the skill set they are looking for, enabling candidates to find roles that align with their expertise and aspirations. If you have a technical background and are interested in product management, the PMT role provides a bridge between their technical skills and their desire to work on product strategy and innovation.

How is PMT different from TPM?

It's not really! Both TPM and PMT have a high degree of overlapping responsibilities and skills. For small programs, a TPM can most likely play the PMT role. This is true for a number of infrastructure-focused TPMs who created technical requirements and worked with internal customers to design and launch systems. There weren't really many Infrastructure PMs until recently. The need for both PMT and TPM on the same program is only beneficial when the systems are highly complex, spanning multiple teams and even organizations.

Example: A highly complex and ambiguous program like autonomous self-driving cars may need all three roles - a PM focused on customer facing experiences like the ride hailing app, a PMT to define autonomous vehicle (AV) behavior, and a TPM to manage programs at various levels like managing perception systems or simulation systems for a new city.

The synergy between these roles ensures that products not only meet customer needs, but also push the boundaries of technical innovation.

Is there a need for a separate PMT role?

No, in my opinion a PMT is just an archetype for a PM - essentially a PM who is more technical, just like a Product TPM is more product-centric. The difference between a TPM and a PMT is the same as between a TPM and a PM.

TPM VS. ENGINEERING MANAGER (EM)

Engineering Managers generally manage one or two teams of engineers. At first glance, the roles of TPM and EM might seem interchangeable, as both involve overseeing technical programs and teams. However, they have their unique contributions to the organization. EM serves as the team's captain, guiding their members through the ever-changing waters of technology. While also responsible for delivery, the EM's primary focus is on the growth and development of their team members. They foster an environment of collaboration, mentorship, and continuous improvement. The Engineering Manager is not just concerned with deadlines; they prioritize cultivating a team culture that encourages technical innovation, learning, and individual growth. As leaders, they understand that empowering their team to thrive ultimately leads to better products and more fulfilled engineers.

TPMs, as we have seen, work across multiple teams and are managing programs and their components. They may partner with multiple EMs to ensure that the right people are working on the right initiatives. They will collaborate with the EMs on deadlines, dependencies, and overall system design. A key goal is ensuring that diverse teams are harmoniously working towards a shared goal. The TPM's role transcends day-to-day technical oversight; they are the architects of structure, ensuring that programs are executed efficiently and successfully.

On smaller programs that are self-contained, EMs can carry the responsibilities of a TPM and can manage individual project schedules and scrum meetings. However, both roles become crucial as programs cut across team boundaries and become more ambiguous in nature. On such occasions, TPMs enable the EM to focus on their team, the deep technical aspects and what they do best - motivate highly technical engineers to build great products in the best possible way.

TPM VS. SCRUM MASTER

A Scrum Master is like a guide for a team using Scrum, an Agile framework commonly used for software development. Their primary job is to ensure that the team follows the Scrum principles and practices, which have specific roles and rituals. The Scrum Master works closely with each team member, coaching and guiding them through the Scrum methodology. They uphold Scrum values and practices while being adaptable and open to improving the team's workflow. A Scrum Master is a facilitator, coach, and sometimes takes on project management tasks, like leading meetings, supporting team members, and handling conflicts (Rehkopf, n.d.) ("What Is a Scrum Master (and How Do I Become One)?" 2023).

On the other hand, a TPM is like an orchestra conductor for the entire program. They oversee multiple teams, projects, and initiatives, aligning them with the organization's goals. While a Scrum Master focuses on the team and Scrum framework, a TPM looks at the bigger picture, ensuring the program aligns with the overall strategy. They navigate complexities, risks, and dependencies across various projects. TPMs communicate with stakeholders and leaders, providing a broader view and often taking on a leading role. In some scenarios, new or junior TPMs handle Scrum Master duties for multiple teams while overseeing the entire program, especially in smaller programs like those in startups. As TPMs advance in their careers, they work closely with engineering leads to manage Scrum activities. Additionally, larger tech companies might use custom or hybrid Scrum frameworks, impacting the visibility of the Scrum Master role.

* * *

In conclusion, not every role is required for every program or team. But every role has its unique strengths that can be levered depending on the circumstances. It is important to understand each role's strengths and align them accordingly for the best outcome.

PRO TIP: The best way to build a strong partnership with a Product Manager or Engineering Manager is to discuss your role, responsibilities, and strengths at the outset. This will help establish clarity, build relationships, and avoid unwanted confusion with responsibilities during the program.

CHAPTER 3

TPM MYTHS DEBUNKED

Is TPM even a job? I have seen a lot of TPM's only doing
some useless meetings and managing sprint dashboards.

– Unknown Blind User

I may have been in this function for more than a decade but it doesn't hurt any less when someone disrespects a TPM or the entire function because of their lack of knowledge or positive experience. This, in part, is what motivated me to share TPMs' duties, and their strengths and value. Eventually, it led me to write this book. I would like to clarify the misconceptions around this role before we dive into how to succeed in Technical Program Management.

The TPM role is still relatively new and with other overlapping roles in the tech industry, it is often misunderstood. Sometimes, even new TPMs may not fully grasp the nuances of the role because they may never have been exposed to them. Due to its broad and fuzzy nature, there is a lack of clarity about what TPMs really do. Many TPMs feel dejected due to a lack of understanding and respect. They often end up moving to other functions, such as Engineering or Product, to experience a deeper sense of identity and belonging. Every role has its strengths that, when leveraged correctly, help elevate the entire organization.

Let's debunk 12 myths so you can build a better understanding of the TPM role.

Myth: TPMs are Project Managers for a product or engineering team

No, TPMs are technical thought leaders and strategic partners to engineering/product teams. They take part in the full product lifecycle from understanding the vision and strategy to brainstorming requirements and technical feasibility, and then launching a high-quality product on time. TPMs do not just build schedules and check things off a list. TPMs utilize their deep domain expertise to build a cohesive and holistic program execution strategy to deliver on business objectives working across cross-functional groups.

Myth: TPMs are essentially launch or release managers

No, TPMs possess a deep understanding of product and solutions to influence prioritization, balance short-term hacks with long-term scalability, assess risks/dependencies, and ensure solid execution with a focus on quality. Of course, many programs have an end goal that can be termed as launch or release. However, a successful launch is the culmination of immense effort made through the entire product development lifecycle.

Myth: TPMs are crisis managers

No, in fact, TPMs should be seen as proactive Crisis *Preventers*. Simply deploying a TPM when a program is already mired in a severe crisis is like calling firefighters after a building has already caught fire. A TPM should be involved from the start of the program, so they can leverage their technical understanding and program expertise to prevent crises and mitigate risks proactively. Putting out program fires can feel very impactful, but true TPM skills help avoid pitfalls and prevent risks from turning into fires.

Myth: TPMs help scale EM or PM

No, TPMs work alongside their engineering/product partners. TPMs should never be brought on by an EM or PM to do the work they don't want to do. TPMs are not note takers or task creators. Bring on TPMs when you have programs that are complex, ambiguous, require multi-quarter technical roadmaps, or are extremely time sensitive.

Myth: TPMs create processes that slow us down

No, in fact, TPMs are like catalysts speeding up teams through carefully formulated lightweight frameworks. These best practices help organizations build highly efficient and effective teams that can scale effortlessly to meet goals and be impactful.

Myth: TPMs setup meetings and take notes

No, TPMs are not calendar schedulers or note takers. While TPMs will set up relevant meetings, take notes pertaining to their program, and follow up on action items, they do so purely to aid the tracking of the program for which they are accountable. Unfortunately, many engineering and product folks ask TPMs to set up their meetings or default them to note taking.

If a meeting doesn't need the TPM to be present or lead it, then it does not need to be scheduled by a TPM. Everyone in the meeting should take turns recording notes and action items. In fact, it is more helpful because TPMs are driving the meeting and trying to keep everyone on track while absorbing the information being presented.

Myth: TPMs are glorified administrators

No, TPMs are not secretaries or administrators, and I mean this with the sincerest appreciation to all the admins out there. By saying TPMs are administrators, both functions are being devalued because they have their own unique strengths and both keep organizations running in their own special way. The key here is to have empathy and respect for both TPMs and administrators. Their job is not easy, whatever you may think. They must leverage each function for their strengths without conflating the two.

Myth: TPMs are not technical

No, as you can see, "technical" is part of the title. In fact, TPMs pride themselves on their technical expertise and judgment. While not all TPMs may code, their ability to understand the domain, see the big picture, connect the dots, and think about long-term scalability and stability is what sets them apart. TPMs can change domains and may need time to ramp up, but that doesn't mean they are not technical. They do not have to design your entire architecture, but they may demonstrate technical thought leadership by asking probing questions.

Encourage their technical learning by sharing resources that you/engineering found helpful. Invite their input on technical decisions and consider their questions with an open mind.

Myth: TPMs are just Scrum Masters who manage sprints

No, TPMs are much more than Scrum Masters. TPMs own and drive end-to-end program strategy and execution working cross-functionally across disciplines to ensure the successful launch of new products/services/systems. Tracking the work done using agile methodology is part of the job but it is not their entire job. New and junior TPMs may start out in a Scrum Master like role, managing in-depth sprint processes. However, for senior and seasoned TPMs, the bigger the program, the higher the chances that TPMs do not run individual team sprints or act like a Scrum Master. Execution is everyone's job and even engineers should know the rituals of agile methodologies. Engineers actually benefit by doing some of their own sprint management because it makes them better at their job of finishing deliverables on time.

Myth: TPMs create tasks for engineers

No, TPMs create tasks that are important for them to track the progress of the program. TPMs will create tasks at the feature or story level, but any sub tasks that are purely engineering in nature do not need to be created by a TPM. TPMs can then create and track dependencies across these higher level tasks, so they can create a timeline and identify critical paths.

Think about this - a TPM creates 100 tasks for 10 engineers versus each engineer creates 10 tasks - which is better? Many granular tasks need implementation details that are best provided by the engineers themselves.

Myth: TPMs are a support function

No, TPMs are strategic partners to engineering and product teams. The trio of functions come together in a complementary fashion to deliver business goals. TPMs are thought leaders who leverage their deep domain expertise to manage complex and cross-functional programs. They drive the program strategy and they are equally accountable for meeting program goals. TPMs are force multipliers who help other functions focus on their core strengths, so better visibility and predictability exists for executive leadership (C-level execs).

Treat the TPMs as equal, ideate with them, invite them to meetings (let the TPM decide if it's relevant or not), ask for their opinion, and make them part of team celebrations. Hold the TPMs accountable by setting mutual expectations. TPMs love to have ownership and thrive in challenging environments.

Myth: TPMs are the same as PMs or PMTs

No, TPMs are Technical Program Managers, which is a completely different function from Product Managers. While they do have some overlapping responsibilities as described in Chapter 2, Product Managers lean more towards vision and strategy (what and why) and TPMs lean more towards execution and delivery (when and how). Different companies have slightly different definitions for both roles. The important thing to keep in mind is that there is a place and time for both roles to coexist in one team or organization.

* * *

PRO TIP: Meet with your key partners and ask them how they would describe the TPM role, whether they have experience working with one and what they would like to see the TPM do. This will help you get a deeper insight into the team's understanding of the role and take appropriate steps as needed.

CHAPTER 4

RISE OF THE TPM

I am willing to give up an engineer
to gain the force multiplier effect of the TPM.

– Engineering Manager

During my time as a TPM organization leader, I was in a meeting with Chris, a senior EM, where I had to break the bad news that I did not have enough TPMs available to staff a large cross-team initiative. This would negatively impact his engineering team. Chris knew that his engineers were spending hours in meetings tracking down dependencies that could otherwise have been used coding for the actual project. He begged me to hire a TPM because he knew the value that the TPM would bring to his team. Chris was willing to give up being an engineer in order to help his team spend its time more productively. TPMs stand shoulder to shoulder with engineering and product peers, allowing them to focus on their core teams while the TPM fortifies the foundation of the program. Chris' actions provide the validation of the impact that a TPM brings to engineering teams yet, questions remain as to why TPMs are needed.

BUSINESS VALUE AND IMPACT

Over the years, in my role as a TPM I implemented various prioritization frameworks - some common ones being Return on

Investment (ROI) versus Effort or Impact, versus Effort. These frameworks helped the organization prioritize which features get done first. Similarly, at a business level I worked with senior leadership (Director, VP, C-level execs) to prioritize which programs get done in this quarter versus next. As companies were trying to build more advanced products, the complexity of implementation went up. However, we were still prioritizing them because the business value (ROI, Impact. etc.) was almost always higher. Most often these were strategic investments that turned into longer term programs. In today's competitive landscape, more companies are taking on strategic bets to differentiate themselves in the market and ensure their relevance and success, even if it takes longer or requires more people. Such high-value initiatives hold the potential for significant impact for both the business landscape and consumers.

Consider the example of building fully autonomous vehicles. The development and deployment of such vehicles have the potential to fundamentally transform our cities and transportation systems. This paradigm shift not only promises a revolutionary change in how we commute, but also presents significant upside for companies that successfully develop and market these autonomous vehicles. The business value derived from such initiatives is immense, with the potential to reshape entire industries and drive substantial financial gains.

Another example is smart devices and wearables, which have revolutionized the way we interact with data and information, making it accessible at our fingertips, through voice commands, and more. Manufacturers of these devices have witnessed tremendous profits through the sale of these products, underscoring the substantial business value generated by such advancements. Moreover, the impact extends beyond financial gains, as these devices have empowered individuals with greater connectivity, convenience, and enhanced lifestyles.

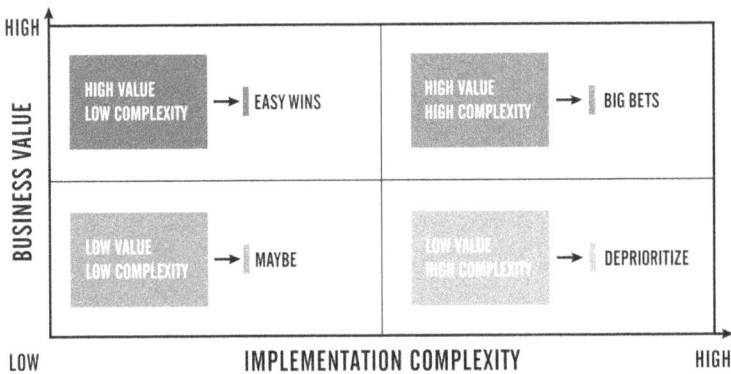

Value-Complexity Matrix

At this point, I hope you are convinced about the correlation between complexity and business value. You may still wonder how TPMs come into the picture. For that, let's look at how uncertainty and complexity are inter-related.

COMPLEXITY AND UNCERTAINTY

Complexity manifests itself in various ways, each posing distinct challenges to programs.

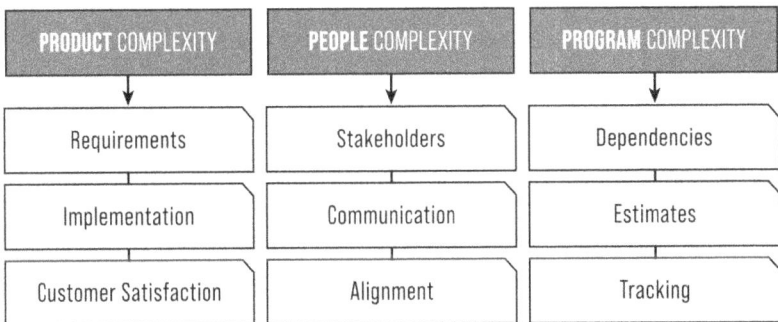

Types of Complexity

Product complexity: Complex products often involve capturing both functional and nonfunctional requirements, which can be intricate and challenging. The technical implementation of such requirements is highly difficult, and ensuring scalability and stability becomes important for the long-term success of the product. Additionally, achieving customer satisfaction becomes trickier when complex products often entail diverse user needs and preferences.

People complexity: Complex products typically require expertise from multiple domains, leading to the involvement of numerous teams and stakeholders. This multiplication of communication lines increases the likelihood of breakdowns and misunderstandings. Aligning the efforts of multiple teams and stakeholders with different perspectives and priorities becomes a complex task, necessitating effective communication and collaboration strategies.

Program complexity: The combined impact of product and people complexity results in program complexity. Tracking dependencies between various components and teams becomes convoluted and difficult. Accurate estimation of timelines and effort becomes more challenging due to the presence of unknown unknowns. Developing a high-confidence roadmap requires iterative planning and continuous adaptation to evolving circumstances.

In the face of high complexity, uncertainties arise regarding technical feasibility, resource availability, market dynamics, and customer preferences. Uncertainty is directly proportional to complexity and vice versa. Either way, the impact on any business is significant. The presence of uncertainty amplifies the risk associated with business initiative. Higher levels of complexity and uncertainty increase the likelihood of failure, translating into higher risk. Minimizing failure and managing risk are crucial objectives for companies to succeed.

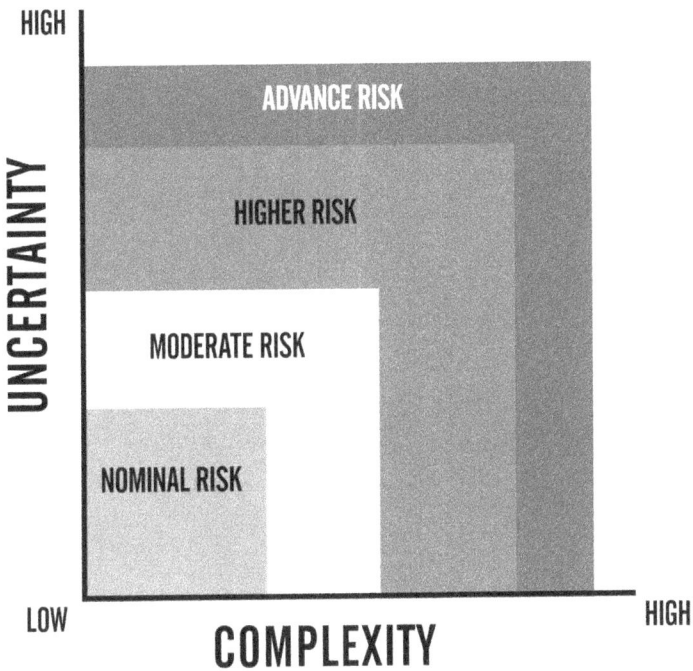

Complexity-Uncertainty-Risk Relationship

The ever-increasing complexity of delivering technological advancements increases the program scope and the need to manage it more effectively. The consequential increase in risk has necessitated the evolution of the TPM role. TPMs possess the skills and expertise necessary to navigate complexity, manage uncertainty, and mitigate risks. They do so by breaking down complex problems into manageable components, devising effective strategies to tackle uncertainties, and implementing risk mitigation measures. By leveraging their technical acumen, program management skills, and strategic thinking, TPMs help lower the chances of failure and ensure successful program outcomes.

Example: A TPM leading the launch of an autonomous vehicle to a new city may need to break the project into known and unknown scopes. For scopes that aren't clear, they may need to set aside a phase for research where they identify the right people and tasks needed to get more clarity. Any scope that lacks clarity is automatically added to a risk register, so the TPM is continuously figuring out a way to mitigate those risks. Additionally, TPMs may need to talk to non-engineering teams regarding compliance, regulatory, or safety related activities that depend on third-parties. This is identified as another risk and tracked accordingly.

With so much at stake in managing complexity, uncertainty, and risk, the TPM role has rightfully claimed its place at the center stage of technical program management, driving organizations towards achieving their strategic goals and delivering impactful results. Over the years, the TPM role has continued to adapt to the needs of an evolving workplace, redefining itself to provide clarity and value.

TPM REDEFINED

In the 10 years since I first started actively defining and clarifying the role for myself and others, there has been abundant change in the business and tech world. Just like rapid technological advancement and transformative changes in various industries has redefined how people work, play, communicate, and connect, many functions and roles have been added, changed, and evolved. The TPM role has also experienced significant redefinition, particularly at top tech companies. Today, being a TPM goes beyond the traditional perception of a project manager, scrum master, task manager, or status reporter.

As companies strive to remain competitive, innovate, and drive growth, the TPM role has emerged as central and indispensable. TPMs act as the driving force behind managing complexity, reducing uncertainty, and unlocking the business value and impact sought by companies. They are getting results through specific actions and using their skills to navigate intricate program landscapes, align stakeholders, and mitigate risks. This positions them as vital contributors to the success of complex initiatives.

They are not just focusing on *what* needs to be done but *how* they are going to do it.

Who Is a TPM?

I first came across the term "entropy crushers" in Michael Lopp's post back in 2013 while trying to create a RACI between TPM and PM. The term resonated with me so much that I used it for many years. It truly represented TPMs and how they manage chaos and bring clarity to the teams around them. TPM is a very action-oriented role, which is why an entropy crusher resonates with me because it captures the proactiveness along with bringing out the human aspect of the person behind the role. On the other hand, words like "glue" or "bridge" that are often used to describe TPMs feel passive and uninspiring. They do not capture the essence and power of what a TPM does. We refer to engineers as builders or inventors or creators. We refer to product managers as visionaries or strategists or pioneers. So why not refer to TPMs with the same fervor?

LET'S REDEFINE TPM → ~~WHAT~~ WHO IS A TPM?

TPMs serve as catalysts for growth, innovation, and efficiency. They contribute to the strategic direction of the organization, drive collaboration, and foster a culture of continuous improvement. By transcending traditional expectations and embracing these broader responsibilities, TPMs play a vital role in shaping the success of their teams and organizations.

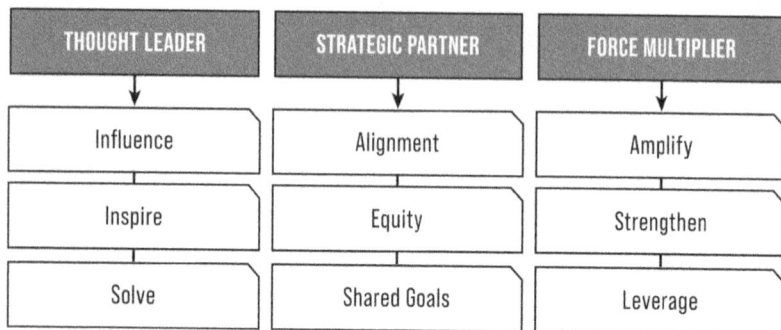

THOUGHT LEADER	STRATEGIC PARTNER	FORCE MULTIPLIER
Influence	Alignment	Amplify
Inspire	Equity	Strengthen
Solve	Shared Goals	Leverage

Who Is a TPM?

Thought Leader

TPMs possess an inherent ability to share their unique perspectives, opinions, and insights in a way that influences the people around them to think outside the box. The TPM is a thought leader who goes beyond merely fulfilling assigned tasks to inspire teams and drive innovation by thinking creatively and encouraging others to do the same. TPMs bring fresh ideas to the table and solve complex problems by challenging the norms and exploring unconventional approaches. We will cover thought leadership in greater detail in Chapter 20.

Strategic Partner

TPMs are in the midst of multiple teams and functions, working with technical and non-technical folks. TPMs partner closely with engineering and product teams on a daily basis, standing shoulder to shoulder. TPMs enable their partners to focus on their core strengths while fortifying the foundation of the program. This approach goes beyond playing a support role. It is a dynamic and equal partnership that drives results. TPMs forge strong connections and foster an open collaborative environment to make decisions together. They cut across organizational boundaries to ensure alignment and establish shared goals to work cohesively towards a common vision. They motivate the team even if it's from the sidelines. That's what partners do - they know when it's time to lead and when it's time to follow.

TPMs have a strategic insight into everything that's going on and that's what they are bringing to the partnerships.

Force Multiplier

TPMs are constantly optimizing and streamlining - whether it's through setting up best practices, frameworks, or automating systems to reduce manual work. TPMs possess the ability to amplify the voice of their organization, advocating for the needs and priorities of their team. They strengthen their team's capabilities and effectiveness by identifying opportunities for improvement, streamlining processes, and removing barriers to success. Additionally, TPMs leverage their own unique skill sets and expertise to bring order to chaos, instill structure, and enhance the productivity of their teams. They are adept at navigating complex environments and optimizing resources to deliver high-impact results.

When a TPM makes every person on the team 10% more effective, that adds up exponentially. As a result, the entire organization becomes more effective, efficient, and productive. That's the force multiplier effect!

As you embark on your journey as a TPM, embrace the redefined nature of the role. Become a thought leader who influences and inspires others, a strategic partner who aligns stakeholders and establishes shared goals, and a force multiplier who amplifies the capabilities and productivity of your teams. By embodying these characteristics and leveraging your unique skill sets, you have the power to make a transformative impact on your organization's trajectory and achieve exceptional results.

* * *

PRO TIP: When anyone asks What is a TPM - Tell them this - It's not *what*, it's *who* is a TPM. Then answer - "I am a -
- **Thought Leader**
- **Strategic Partner**
- **Force Multiplier**
- *Entropy Crusher*
- *Catalyst*
- *Problem Solver*
- *Doer*
- *Connector*
- *And much more*

CHAPTER 5

CORE TPM SKILLS

Without Knowledge, Skill cannot be focused.
Without Skill, Strength cannot be brought to bear.
Without Strength, Knowledge may not be applied.

– Alexander the Great

In my first formal TPM role at a startup, there weren't any clearly defined expectations or skills to get to the next level. The goal was to deliver multiple programs on time with high quality. I knew that my technical skills were helpful to work with engineers and that I needed to know how to plan, break down the tasks, manage them in Jira, and create an effective execution system. While I used to be an engineer, working with strong-headed Silicon Valley engineers was still challenging for me as an introverted and timid person. My then-manager Ann encouraged me to stand up for myself, lean into my execution strengths, and showed me how best to go from planning to execution to delivery. As one of the first TPMs focused on product launches, I grew with the team, eventually mentoring many of them. However, it wasn't until I started working at a big tech company that I saw a much more structured way of defining the expectations (skills, behaviors, technical acumen, etc.) required for TPMs at each level. The clarity really helped me to continue to elevate my craft and hone my skills.

Through the years, I witnessed the multifaceted nature of the TPM role. The diverse skill sets can be categorized in three key areas or along three main axes derived from the letters in the title - T, P and M. That's T for technical expertise, P for program execution skills, and M for leadership management abilities. To illustrate this concept, envision a monument that embodies impact and success standing atop the pillars of the core TPM skills. At its base lies a solid technical foundation, but to reach the pinnacle of success, one must also reinforce the key pillars of communication, leadership, and program management skills. Becoming a well-rounded TPM involves strengthening each of these pillars in tandem, as they collectively support the achievement of impactful results in complex and dynamic program environments.

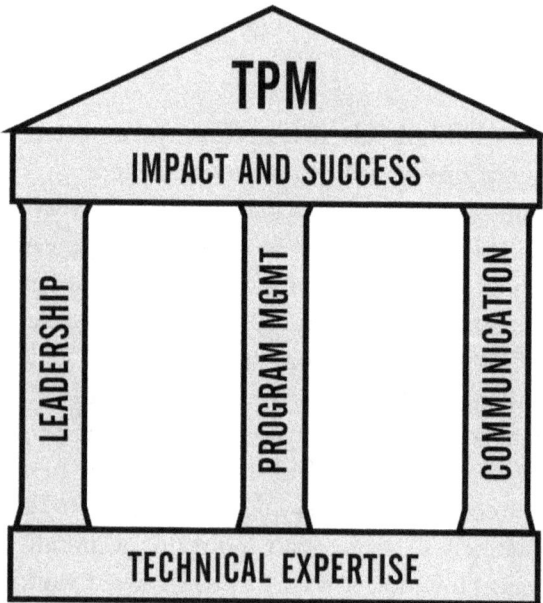

TPM Success Monument

At a high level, you can outline TPM responsibilities as follows:

- Own and drive end-to-end program strategy and execution working cross-functionally around disciplines to ensure successful launch of new products.

- Leverage strong program sense to drive strategic execution by breaking down complexity, assessing priorities, identifying risks, and driving issues to resolution quickly.

- Communicate early and often, utilizing a data driven approach with a focus on the audience.

- Exemplify product and technical thought leadership to identify scalable long-term solutions.

- Influence teams to build holistic and consistent product experience for the end user.

TECHNICAL EXPERTISE

Technical expertise forms the bedrock of success for a TPM. While an engineering background can be advantageous, it's not a prerequisite for developing technical proficiency in this role.

Coding skills are not required, however possessing a deep understanding of technical domains empowers TPMs to make valuable contributions to programs and engage in meaningful discussions with engineering teams. This technical proficiency aids in developing sound judgment for system designs and solutions, and enabling better comprehension of engineering complexities like estimating effort, identifying dependencies, and assessing risks.

Numerous avenues are available to build and enhance technical knowledge, ensuring that TPMs can effectively navigate complex programs and collaborate with engineering teams.

To build technical expertise, you can explore a variety of educational resources, such as courses, certifications, and training programs (detailed in later chapters). For those entering unfamiliar domains, seeking guidance from subject matter experts can be immensely beneficial. Learning from experienced individuals who possess deep knowledge in the field can provide valuable insights and accelerate the understanding of technical concepts. I encourage you to buddy up with an engineer and ask them to explain the technical architecture. It will not only help you grasp the complex technical concepts faster, but it will also help you build strong relationships.

Given the rapid pace of technological change, curiosity is a valuable trait for TPMs. Staying inquisitive and regularly updating your technical foundation is essential to remaining relevant. By building and sustaining technical expertise, you can confidently contribute to programs, offer valuable insights, and facilitate effective communication between technical teams and non-technical stakeholders. Engineers are also more likely to resonate with your feedback, accept suggestions and respect your opinion if they feel you understand the technical complexity.

PROGRAM MANAGEMENT

Program management is the bread and butter of the TPM role. It involves overseeing a collection of interrelated projects and managing ongoing iterative changes. Program frameworks help with decision-making and improve delivery, ultimately achieving desired business goals for the team. However, I want to emphasize that program management skills go beyond having the knowledge of tools like Jira or methodologies like Agile and Waterfall. It's important not to get bogged down in the details of sprint boards or tasks, but to focus on eliminating risks and reducing complexity. The tools and methodologies are merely a means to an end. Effective program management includes proactive planning, effective prioritization, effortless tracking, and flawless risk management.

Let's work through the skills required to handle a core program management responsibility like use product requirements, business needs, and technical challenges to develop program schedules. When you see a broad statement like this, think about all the actions needed before you can finalize a program schedule or deliver on a goal. Your role is not just about creating a schedule and beautiful Gantt charts with every dependency and milestone laid out. An experienced TPM knows that a program tracker is the outcome of all the hard work that goes on behind the scenes. It's merely a communication tool rather than the core of their work, which happens throughout the program management lifecycle.

Planning and Kickoff

In order to develop a realistic schedule, you have to understand the problem space, business objectives, and desired outcomes. You should be able to take a complex and ambiguous problem and break it down into manageable pieces. Once the business objective is clear, the TPM needs to identify key stakeholders (like product, engineering, testing, privacy, security, etc.).

You will drive discussions with multiple teams to gain clarity and align everyone with shared goals and proposed solutions. You will use your domain expertise to influence requirements, define success criteria, and review technical designs.

Example: Let's take Google Maps as a program case. The business objective may say "Launch Google Maps." As a TPM, you will need to understand the Maps domain and provide context. Is this a brand new functionality, or are you launching in a new locale or country?

Prioritization

In order to start the development phase, you will need to go through a prioritization exercise. Your business goal may determine the factors that go into the prioritization. You will need to finalize the scope and identify phases and milestones. Your technical/product know-how will help you identify dependencies, suggest alternatives, and propose scalable future-proof solutions. You gather effort estimates, sequence the work, and create a high level roadmap and timeline. As a TPM, you also need to ensure that conflicting priorities for different teams are taken into account, so that dependencies can be fulfilled in a timely manner.

Example: *If launching Maps in a new country, does the scope involve providing basic maps and driving directions? Do we also launch traffic information or integrate local listings? Is mapping and calibration a dependency for the traffic team? Do we need to understand country specific privacy/legal regulations? Will our architecture support x million more users if we add a new country? Can we launch in the next country faster without refactoring code?*

Execution and Tracking

A majority of the development work happens during this phase. This development phase is not limited to software only and can be applied to any programs, including hardware. As a TPM, you will take the high level timeline and produce a granular work breakdown structure. You will decide on the execution strategy - methodology, tools, etc. You will now work with people assigned to tasks and initiatives, track progress, and unblock individuals to ensure the program is proceeding smoothly. You will start incorporating testing and validation activities into this phase.

Risk Management

Risk assessment happens through all phases of a program and is one of the most critical program management skills that all TPMs must possess. Risks can cause a program to be delayed or even fail, which can lead to loss of money, users, reputation, etc. Risk mitigation includes identifying and classifying risks by understanding historical context and probability of failure based on what is known versus unknown. Once you have risks, you need to create a plan to effectively mitigate the risks, so that the program can still be launched on time. We will dive deeper into risk mitigation in Chapter 17.

Reporting and Communication

This is essential to keep all program stakeholders informed about the progress, risks, timeline, etc. As a TPM, you need to create a tiered communication plan, align to the audience, escalate early and often, make it predictable, and adjust the strategy as needed.

Post-launch

Even after the program meets its initial objective, there are numerous activities involved in the post-launch phase such as setting up support, measuring success through metrics, and conducting retrospectives to identify lessons learned.

PRO TIP: Whenever you report a risk, make sure to include a mitigation plan with details such as mitigation date, name of responsible individual, and impact the customer. Having these details provides reassurance to stakeholders that you have a clear path forward.

I encourage you to leverage additional program management resources (certifications, books, etc.) to dive deep into the phases if you are not familiar with different program management frameworks, tools, and systems.

COMMUNICATION

Communication can make or break the reputation of a TPM. While communication skills are important for all professionals, good communication skills are extremely important to becoming a successful TPM. Effective communication enables you to articulate ideas persuasively, gain buy-in for initiatives, and foster a collaborative environment. This helps address conflicts, overcome challenges, and drive successful outcomes in programs. It is a key enabler for establishing trust, inspiring confidence, and becoming a respected leader within the organization.

Many years ago, I needed to request information from leaders across the organization. They were partner teams, and my team was making changes to our service model, which required this information. After a discussion with senior leaders in my team, I composed the email and sent it out. It was a total disaster! Those leaders were caught off guard with the request and the associated deadline. They had no idea why I was asking them for that specific information. The senior leaders and I had to do some damage control, and it impacted my credibility. The mistake I made was not providing proper context regarding the request and the upcoming changes. I had assumed that all senior leaders had already discussed this prior to my sending the email. So I focused on the action of sending the email with the request and overlooked the background and emotions behind the actions. Communication is the skill I have focused on the most throughout my career and continue to do so, because I think there is always more to learn.

Communication is also challenging because your circumstances are constantly changing. Your organization culture, audience's needs, and program phase all determine the impact of your communication. Impactful communication encompasses much more than mere written or verbal exchanges. It revolves around the art of understanding one's audience, their preferences, and expertly tailoring communication styles to meet their expectations.

Example: Producing a status update for programs is a common form of written communication from TPMs. I have seen hundreds of status reports in my career. The best ones are clear and concise. They provide critical insights that are important for the senior leaders along with specific information that answers the key questions. The worst ones raise alarm bells without providing answers or are a verbatim copy of meeting notes.

A robust communication strategy is also integral to the program execution strategy. The goal of program communication is two-fold. First, to keep the core program team aligned on scope, milestones, and timeline, and second, to keep leadership and stakeholders informed of progress, risks, and mitigations. As the TPM, you need to bubble up the right level of information at the right time to get the most out of your messaging. Besides informing stakeholders, the goal of communication is also to seek help if needed for further execution. You need to identify the right communication strategy (*cadence, level of details, documentation, reviews, etc.*) depending on the audience and the type of information being communicated.

I want to clarify that language proficiency is a distinct skill from communication. Even if you are proficient in a particular language, remain mindful of your communication style and ensure that the intended message is conveyed clearly and appropriately. Language barriers, jargon, or technical terms should be carefully navigated to ensure that the essence of the message is not lost.

PRO TIP: Don't view communication as just an exchange of information. Focus on fostering a deep connection through thoughtful and impactful conversations.

LEADERSHIP AND STAKEHOLDER MANAGEMENT

Stan, a TPM on my team, was an amazing human and highly technical person who knew everything about system architecture. He was also good at planning and creating a work breakdown structure. He could facilitate a meeting and track the progress of tasks in the system. Unfortunately, he was unable to level up and manage his scope because he could not keep up with the work across multiple teams. He was spending too much time down in the weeds trying to manage tasks that should have been done by the team. This meant he did not have enough time to follow up on dependencies, anticipate the risks on time, or build the right relationships. While he was maintaining a dashboard and sending daily emails, he was not communicating clearly and providing the information the leadership team needed. His inability to influence teams or articulate clearly about the issues often led to escalations.

It was very difficult for me to see him struggle even after continuous mentorship. He kept doing the things that worked when managing one small project with 10 engineers, but could not scale up to five teams with 50+ engineers.

While technical knowledge and program management skills are weighed more heavily for entry-level TPM roles, leadership skills decide your growth in the ranks. As you progress in your TPM career, you will have to lean into your soft skills more often. I truly believe that within these leadership skills is where lies *The Art of Strategic Execution*. It's common to see highly technical TPMs like Stan struggle to advance beyond the junior TPM level because they focused exclusively on expanding their system knowledge and perfecting the tactics or science of execution.

The increased cross-functional aspects of delivering a complex program combined with a distributed and hybrid workforce means that leadership skills play an increasingly critical role. Most TPMs work in a matrix organizational environment, which means that people and leadership skills are a must to be successful. Soft skills are quite important tools that a TPM should have in their repertoire.

One example of a TPM responsibility is to create best practices and product development processes while operating in a fast-paced and dynamic environment. At first glance, it feels like TPM is responsible for developing processes and best practices. However, bringing about a change through these processes is one of the most challenging aspects of the job. It is not enough to write up a document that outlines a new process. The real impact only comes when the process is agreed upon and adopted by all relevant stakeholders. This situation demands a high degree of leadership and stakeholder management skills.

During the planning phase, you identify those key stakeholders (core team, partner teams, managers, etc.). These folks can be internal to your team, cross-functional (outside your main team), or external to the company. As a TPM, you will have to identify the right strategy to work with each of these groups to ensure that interteam and/or organizational dependencies are successfully resolved. You will need to learn more about each member of your core program team to understand their motivations, collaboration style, etc. Change is hard for everyone, so it is important to empathize and understand the viewpoint of your stakeholders. If you want to implement a new execution structure, the long-term benefits like improved visibility and efficiency may not be perceived similarly by engineers. In such a case, you need to work with multiple influential peers, address their pain points, and aim to create a win-win situation. You will need to influence without formal authority, manage the team/partners to foster alignment, and resolve conflicts objectively. Be agile and open-minded during your conversations with stakeholders, and be quick to adapt based on feedback that you receive. This approach of openness and trust will make it easier to motivate the team and solve problems together.

At times, you may be in a situation that demands negotiation or escalation to address dependencies or priorities across teams. It is important for a TPM to take an objective and data driven approach to create alignment and get the desired outcome. You will need to bring in all the data to support your proposal and ask for the same from the other team, so you can both look at it objectively. You will work together to propose the right solutions to manage bottlenecks and quality issues.

The TPM role is extremely people-oriented, and smart, effective collaboration is key to achieving your program goals. The impact you achieve is amplified with the right behavior, which often involves working with people in a constructive, respectful, and inclusive manner. Think deeply into what actions would drive the outcomes required of you. What lessons have you learned as you developed strategies to address challenges in the areas of program planning, communication, or collaboration?

* * *

PRO TIP: A TPM's role is not just about setting up program structure or sending regular status updates. Focus on positive results and how you achieved them, and then refine them.

CHAPTER 6

IS THIS THE RIGHT ROLE FOR YOU?

The worst thing about the TPM role is that it's fuzzy.
The great thing about the TPM role is that it's fuzzy!

– Yours truly

During a TPM panel event, everyone was asked how they came into the role of being a TPM and how they navigated their career through the years. Each panelist had stories about how they chanced upon the role - either through circumstance or happenstance. I was the only one who had clearly charted out a path to TPM and planned every career move. I told them how I took advantage of the fuzzy nature of the TPM role and defined what I wanted to do, and I actually enjoyed not being boxed in. However, not having a clear definition has its pros and cons, depending on your approach. Due to the manner in which the TPM role came into existence, there can still be confusion about the responsibilities of TPMs. The earlier project or program managers were mostly responsible for checking things off their list and had the lever of budget and resources to deliver a defined scope. In the software industry, the same norms and culture weren't present, which meant that PgMs in this space did not have top-down mandate or control over the teams. Thus came about the caricature of a project manager with a clipboard even with the advent of the TPM role.

The exact responsibilities and expectations can differ from company to company, or even from team to team. How much do TPMs get involved in technical decisions? How do TPMs track dependencies? What is their responsibility if an EM or PM is present and many such questions often come up? This confusion leads to misaligned expectations between TPMs and their stakeholders. As mentioned in Chapter 1, a project manager's role is finite and well-defined whereas a TPM's role is up for interpretation and depends on various factors. For that reason, TPMs themselves have to make an additional effort to understand their responsibilities, align to the needs of the program, and define their role.

If you are contemplating the prospect of a career as a TPM, it is important to explore the role's essence and its compatibility with your inherent strengths. TPMs are integral to managing intricate technical programs, which makes it an aspirational path for many and especially those who want to stay close to technology but may not want to code.

The TPM journey comes with both rewards and challenges, but it's not for everyone. It calls for a unique mix of technical expertise, program management skills, and leadership abilities. You should have a strong interest in technology and a determination to drive change to succeed in this role. Since TPMs work at the intersection of technology and people, good communication and interpersonal skills are crucial. If you're more introverted, this aspect might be a bit challenging. The role can be ambiguous, and you'll often need to advocate for its importance. It can be exhausting for some, and requires genuine passion to navigate. TPMs usually work behind the scenes without full control over outcomes or immediate recognition. You need to be comfortable with these factors to truly enjoy and excel in the role.

Here are four actions you can take to nail down your decision on whether to make the move to this role.

Articulate the "Why": What are Your Motivations?

As you think about becoming a TPM, perform an introspection to understand more about your motivations. What propels you towards this role? Which aspects of the role excite you? At the same time, which aspects of the role might pose challenges? Have you considered potential pitfalls that might come to light? Addressing these queries will help clarify your goals. The grass is always greener on the other side, so it's important to differentiate between genuine desire and superficial incentives. This introspective exercise serves to crystalize your motivations, while concurrently proving to be an invaluable asset during interviews.

Leverage Your Strengths

Align your strengths to increase longevity in the role. How do your inherent talents synchronize with the demands of the TPM role? Recognizing your strengths and understanding if they align with the responsibilities at hand is important to succeed. This alignment guarantees that your innate qualities serve as instrumental resources in your journey as a TPM.

Example: You are highly organized and very detail-oriented. These traits can be a strength as a TPM, making it easy for you to execute and track programs.

Define Your Boundaries

Delineating your non-negotiables - what do you want to preserve as you venture into the role of a TPM? Acknowledging these boundaries empowers you to navigate decisions in accordance with your core values.

Example: If you want to avoid confrontations, then the TPM role may not be ideal for you.

Solidify Your Commitment

Commitment frequently requires sacrifice. What are you prepared to relinquish in order to transcend your inhibitions and flourish in the TPM role? Contemplating this question imparts insights into your resolve, passion, and the fortitude you bring to the table.

Example: If you are an engineer wanting to become a TPM, you will have to give up direct control over the implementation details once you transition over. While you will still have the ability to influence engineering, you cannot actually go in and code even if you may know how.

* * *

PRO TIP: Write down your thoughts to truly synthesize what you are looking for in the next role. Writing helps the brain process information differently than just thinking it over in your head.

TPM SPOTLIGHT

MICHAEL GÖTZ, TECHNICAL PROGRAM MANAGER LEADER

Michael started his career as an aerospace engineer, but soon realized the aerospace industry lacked the speed and innovation he was seeking. Michael ventured into the growing startup scene in Munich while pursuing an MBA in parallel. He co-founded two startups in the IT/software industry in Munich. While only one of them was successful, he gained valuable experience that eventually led him to join a company specialized in process mining and execution management. Michael served as a Product Manager, focused on analytics and data visualization. He was part of the team that launched the first cloud SaaS product for process mining. He saw the company grow from 100 to 3,000 employees over the course of six years.

How did your journey lead you to become a TPM Leader?

Michael: As my company expanded, I worked closely with the CTO/Founder for two years, handling various dimensions such as scaling and growth for engineering, overseeing operations, helping establish information security standards, and setting up a Program Management Office (PMO) to support the product and engineering organization comprising 150-200 people. We also underwent several successful acquisitions, where I was involved in the tech and product due diligence.

During this period, I met the new Chief Engineering Officer and proposed the creation of the TPM role to support the growing engineering organization and to address the need for more technical expertise within the PMO. I transitioned from a PMO lead role to a TPM role to better focus on my responsibilities. I focused on creating a TPM career path, putting together knowledge bases, growing a TPM team, and listing out certifications that would help with growth and development of the team. Today, I drive the organizational setup (career path, job description, learning and development) besides leading a team that is one part of our global TPM organization. Our TPM Leads are decentralized and embedded with engineering organizations aligned to the SVPs. All the TPMs have software engineering expertise or have worked in that environment.

How did you set up the TPM career path at your company, and what factors did you consider during its design?

Michael: Setting up the TPM career path was a detailed process. I worked closely with HR and conducted extensive research, seeking materials and insights from other TPM Leaders. I initiated meetings with SVPs and Engineering Managers to understand the value and impact of TPMs to our company, and to determine the initial focus with TPMs driving strategic programs across engineering teams.

The TPM career path was designed around several key axes, including scope and impact, collaboration, complexity and ambiguity, and, foremost, technical acumen. These criteria increase with each level of the TPM career path, which includes upwards the ladder - Associate TPM, TPM, Senior TPM, Staff TPM, and Principal TPM.

At every level, TPMs are responsible for breaking down work into actionable packages, milestones, and ultimately delivering fully functional solutions as part of the program. Important to note: TPMs each have 3-5 programs. However, each TPM owns and drives a lighthouse program while growing their domain expertise and working across teams to track dependencies and support planning.

The career ladder focuses on three clusters:

1. A general work area profile that determines complexity and ambiguity of the programs for each TPM level.

2. Program Execution and technical acumen that outlines how TPMs manage programs, collaborate across teams, and handle change management.

3. TPMs have specific expectations and responsibilities at each level, ensuring clarity and consistency in their roles including strategic work, emphasizing communication, teamwork, leadership, cultural values, and personal growth.

The role expectations are published across the organization for all TPMs to ensure consistency, and we also created a hiring handbook to ensure consistency when hiring TPMs in different areas.

How do you guide TPMs to grow along this career path?

Michael: We have performance management, which involves self-reflection and growth planning, with managers providing feedback and aligning on individual development needs. We have regular career development conversations every six weeks to help ensure TPMs' growth and alignment with the company's goals.

How can TPMs be best leveraged within an organization?

Michael: To effectively leverage TPMs, several key factors come into play. First, TPMs should be actively involved in the planning process, ensuring they are aware of the organization's goals and objectives. They should collaborate and connect closely with Engineering Managers on programs that require cross-team coordination, leveraging the technical expertise of TPMs to ramp up quickly on the domain.

It's important not to overload TPMs with too many programs. Instead, assign them a single, complex program and a couple of smaller ones to maintain focus and effectiveness. Additionally, having C-level sponsorship is crucial for the TPM and the program success, and it requires alignment among leaders on the responsibilities and value TPMs bring to the organization. TPMs should also take the initiative to educate their partners about the value they can bring, ultimately becoming indispensable to the teams they work with.

What advice would you give to other TPMs or aspiring TPMs?

Michael: Building strong relationships with Engineering Managers is essential. Establishing a broader network outside of your immediate program team and company can also be valuable for professional growth. Remember that it's better to excel in managing one program exceptionally well than to handle multiple programs with mediocre results. Communication is key, and TPMs should never hesitate to ask for help or raise their concerns early and positively escalate issues when needed.

PUT IT INTO PRACTICE

ASPIRING TO BECOME A TPM?

✓ Talk to TPMs in your organization, shadow them for a week, and get their perspective.

✓ Create a list of programs where you may already be acting as a TPM and reflect on your experience.

✓ Assess your current skills and identify areas where you need to build certain core skills.

ALREADY A TPM?

✓ Connect with your EM and PM partners and talk about each of your roles, responsibilities, and strengths.

✓ Shift your mindset to adopt the avatars - Thought Leader, Strategic Partner, Force Multiplier.

✓ Share the myths about the TPM role with someone who may have never worked with one.

Part 2

LAUNCHING YOUR TPM CAREER

Every great dream begins with a dreamer. Always remember,
you have within you the strength, the patience,
and the passion to reach for the stars to change the world.

– Harriet Tubman

CHAPTER 7

UNDERSTANDING YOUR OPTIONS

If an opportunity doesn't knock, build a door.

– Milton Berle

I had the pleasure of meeting Dana, a new TPM who was eager to grow in her role and make a difference. She told me she used to be a dental hygienist for five years before deciding to shift into tech a couple of years ago. Dana no longer wanted to sit in a dental office, doing the same thing over and over. She wanted to feel challenged and decided to get trained as a software engineer. Working at a startup gave Dana opportunities to use her execution and stakeholder management strengths in addition to the newly acquired tech skills, which led her to her current role as a TPM. I was so inspired and amazed by the career pivot she had made. We may often think that there are only a couple of ways to get into a particular role, but Dana's story is proof that there is no one way; there are multiple paths to find the role you love.

PATH TO TPM

The beauty of the TPM role is its versatility, which is also reflected in the backgrounds of so many wonderful TPMs. While the early TPMs, including me, came from engineering roles, having an engineering degree or a prior engineer title is no longer a requirement. There are many different ways to enter the TPM field. Many TPMs have prior program or project management experience in other fields that may not necessarily be technical. These folks are building technical expertise in other non-traditional ways. Many TPMs have product, design, or operations backgrounds. There are some TPMs that come from completely unrelated industries or fields like Dana.

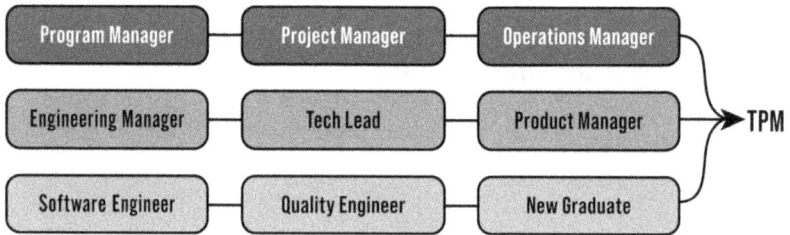

Path to TPM

Most often, TPMs come with 3-5 years of prior experience in other roles, which is helpful for them to navigate the requirements of operating across teams in a work setting. In the last 2-3 years, there has been some uptick in TPM internships or entry level roles. However, with the complexity of the TPM responsibilities, it is best to spend the early years understanding the product development lifecycle from a different perspective before moving to TPM.

Whatever the path, a good understanding of the technical systems and program management lifecycle is essential to making a successful transition to TPM.

TPM ARCHETYPES

I like to say - TPMs are like ice cream. They come in many flavors and each flavor is unique and delicious in its own way. TPMs bring different strengths and are great at adapting to the unique needs of the different teams they work with. Their role and responsibility can also shift based on the specific domains they operate in. The best aspect of this role is its malleability—no two TPM roles are identical. TPMs embody a rich variety of archetypes, each tailored to address particular organizational requirements.

The four TPM archetypes most commonly seen across various organizations are:

- Generalist
- Specialist
- Product
- Infra

In my initial days as a TPM, I worked on various programs - product, backend systems, infrastructure, etc. I found I always enjoyed working on user facing products and therefore took on the Product TPM archetype. Over the years, my role evolved to be more breadth focused, similar to a generalist, working across both product and infrastructure teams.

Here's a brief overview of each TPM archetype.

Generalist TPMs

Characterized by their adaptability across different domains, generalist TPMs possess strong program management skills that can apply to various projects or initiatives, regardless of the specific technical details. Generalist TPMs might not dive as deeply into the technical intricacies as other archetypes, but they are versatile in their ability to switch between different programs and effectively manage them. Their value lies in their agility and capacity to navigate diverse landscapes with a broad program management skill set.

They oversee large-scale programs from beginning to end, encompassing a wide range of tasks and projects. They often excel in managing complexity by understanding the holistic view of the program and how different pieces fit together. They may collaborate closely with other TPMs, aligning individual efforts and ensuring the overall success of the program.

They may be in charge of launching major products or services. Their role involves coordinating various teams, resources, and timelines to ensure that all components of the program are integrated seamlessly and that the launch occurs on schedule. They may be entrusted with the critical task of aligning hardware, software, and other elements to create a cohesive and functional whole.

Example: A TPM is responsible for launching a device like a Virtual Reality headset. They shoulder the immense responsibility of ensuring all hardware and software components align perfectly to meet deadlines. They will work with other TPMs who handle specific components or programs that feed into the headset - like controls, screen, etc. However, they don't dive deep into those systems and instead focus on ensuring all work streams are tracking towards a launch deadline. They still understand the overall system and how different parts fit together in order to create an effective program strategy.

Specialist TPMs

These TPMs possess multi-year experience in a specific domain, which might encompass various technical and non-technical aspects. They bring a holistic understanding of how their domain functions within the broader context of the organization. They excel in aligning the technical requirements of their domain with the overall goals of the program and its future. TPMs are often subject matter experts within a specific technical domain. They immerse themselves deeply into complex and intricate areas, and excel in understanding the nuances of the technologies, processes, or systems they are managing. They often work with other highly technical subject matter experts. TPMs often oversee specific elements or components of a larger program or project. Their role is to ensure that the assigned component is developed, integrated, and managed effectively within the broader context of the program.

Example: They might specialize in AI and ML technology or they may be responsible for the camera component of the virtual reality headset device.

Infra TPMs

TPMs who manage the backend systems, services, and infrastructure that support an organization's operations, fall into the Infra TPM category. Their core customers are generally internal to the company. Their role involves ensuring the stability, scalability, and efficiency of underlying technical systems, which may include databases, servers, networking, and other critical components upon which thousands of other products or services may run. Infra TPMs bring deep technical expertise of components (hardware or software) to collaborate with technical teams, and plan and execute programs that enhance or maintain the organization's infrastructure. Their success is often measured by factors like system uptime, and performance optimization. They also may work closely with product teams to understand how new products impact underlying infrastructure - whether new systems have to be built, or existing systems need to be updated to handle new data or load. In some companies or programs, Infra TPM may also write the technical or product requirements, or both, for the technical systems.

Example: *An Infra TPM would work on projects that improve the stability and performance of a machine learning infrastructure or they might develop new data infrastructure capabilities to support AI products.*

PRODUCT TPMS

Product TPMs demonstrate a keen product sense and specialize in managing the development and delivery of products directly catering to end users. They work closely with cross-functional teams, especially in product and engineering, to ensure that the product's features, functionality, and usability align with user requirements. Overseeing the entire product lifecycle, from conception to launch and beyond, their focus is on delivering value to the end user, with success measured by how well the product meets user needs and drives positive user experiences. Both Product TPMs and Product Managers can exist on the same program and work together to make the program successful.

Example: A Product TPM might oversee the development and launch of a new mobile application, understanding how users will use the app's features and how usability can be improved for higher retention and growth.

The Product TPM archetype came into formal existence around 2016 as a separate job profile for TPMs who worked with product teams that were developing customer-centric experiences. The need for product sense, problem definition, and customer empathy led to the differentiation between product and Infra TPMs. However, many companies always had TPMs working on user facing products without differentiating them from other TPM archetypes.

* * *

PRO TIP: Don't worry about choosing an archetype in the early stages of being a TPM. Work on different products and programs and switch between different archetypes. This will help you understand your preferred archetype, which will be based on your strengths, skill sets, and interests.

CHAPTER 8

PLANNING YOUR TRANSITION

Nobody ever mastered any skill except through intensive persistent and intelligent practice.

– Norman Vincent Peale

The first step in managing any program is to make a plan. Planning increases the chances of success. Similarly, transitioning into a different role requires intentionality and planning. It took me more than two years to officially transition from an engineer to a TPM. I was already a technical lead and had managed a couple of projects from start to finish. I gathered resources, got a business degree, and kept asking for more projects and also built out a framework for handling production issues. I was able to use this experience on my resume and in interviews. More importantly, because I had managed these programs, I knew how to create a plan, manage risks, and find workarounds. This helped me provide specific examples during interviews. It also was helpful because when I landed the TPM role, I could ramp up quickly and work with little guidance.

As you plan your move, start building out a diverse set of experiences and assess which core TPM skills you need to strengthen and where you can leverage your strengths. Depending on where you are in your career journey and your background, your plan may look different. If you have a technical background, build your program management skills by taking on some team projects. If you do not have a technical background, start by building a strong foundation in the areas where you already have familiarity. If you work in a small team, take on bigger programs that give you the opportunity to work with multiple teams and hone your stakeholder management skills. Here are five tips to start developing the experience without having an official title.

Get Involved in the Planning Process

TPMs are responsible for planning and executing complex programs. To gain experience in this area, get involved in the planning process for your team. It will give you an opportunity to understand how requirements are captured, how teams do capacity planning, and how high-level requirements turn into execution plans. Pay special attention to how risks are identified during the planning process. Take part in tradeoff discussions and what can help mitigate those risks. Depending on your organization's planning cadence, you will get exposure to both product and business strategy, which will help you build a broader view of the organization, cross-dependencies, and priorities. Shadow a PM or TPM on your team, observe their work, and ask lots of questions.

Start Managing Smaller Projects or Initiatives

Look for opportunities to manage projects in your current role or volunteer to share responsibilities with another TPM. This will help you put on your TPM hat and start thinking about your projects from a TPM perspective.

Example: *Think about the risks that may come up in the project and what can be done to mitigate it. Or, consider what the current gaps or dependencies are that you know or need clarity. Even if the risk is not on your deliverable, discuss your thoughts with the existing TPM and ask what they think. Check with them on how the project is being tracked and discuss any specific tools or strategies that you can use in the future.*

Communicate Updates Regularly

TPMs have to routinely report on the progress of their programs. This requires your communication skills to be top-notch. If you are not already doing so, start communicating and sharing updates about your work. Regularly crafting communication messages for different stakeholders will help you develop that skill and muscle. Learn how the audience consumes information and what is important to them. I have seen many senior engineers do this well, so this is a skill that is required universally at senior levels.

Find Presentation Opportunities

To further build on your communication, hone your presentation skills and build visibility with senior stakeholders. There will be numerous occasions where you will find opportunities to present and build influence, negotiate resources, or convince someone of your ideas. The more you present, the better you will get at understanding where you need to level up and what senior stakeholders look for during such presentations. Monthly execution reviews or weekly risk reviews can be good avenues, which can then evolve to presenting at All Hands types of events.

For me, joining a group like Toastmasters was helpful in building confidence to present. I also took additional presentation skills classes to refine my public speaking style. So, find opportunities at work or outside of it.

Network with the TPM Community

Learning from other TPMs can be a valuable way to gain experience and learn more about the role. Observing TPMs in their work will help you understand how good TPMs make it seem so effortless. Attend industry events, join online communities, or seek mentors who can provide guidance and advice. Look for opportunities to present at community events and make attending learning events a priority. I know one can get busy with day-to-day work, but here's where I want to emphasize that if you are serious about transitioning to the role; you need to prioritize these types of opportunities.

DEVELOPING THE T IN TPM

When I started working formally as a TPM at a small AdTech startup, I didn't feel the need to prove my technical skills because I thought everyone knew I had an engineering background. I was working with them on various programs that were highly technical. When another male TPM joined the team after a few months, everyone, including the Vice President (VP) of engineering, always thought that he was more technical than me. Then came along a hackathon event where I led an initiative to create a tool that automated the generation of the company's launch roadmap, which was being produced manually every week in Excel! I roped in other engineers and even though they knew this was not going to win any awards; I sold them my vision. They took my requirements and built an Minimum Viable Product (MVP) version that saved us TPMs 10 hours each month. After the hackathon, I maintained and enhanced the tool, adding new features by checking in code into Github. I didn't know the significance of this moment until I went to get my noise cancellation headphones, which were given to the engineering team to help with concentration. A senior engineer handed them to me, saying, "You deserve them because you are checking in code." A few hundred lines of code was bigger proof of my technical skills than the previous seven years as a software engineer or earning two engineering degrees.

As you can see, my technical skills not only differentiated me from other TPMs but also built credibility with my engineering partners. The "T" in TPM is crucial, differentiating TPMs from other types of program managers. TPMs work with subject matter experts daily and it is important for them to understand and converse with expert technical folks easily. TPMs drive technical execution, which will suffer if a TPM is lacking on the technical axis.

If you come from an engineering background, it is often easier to transition into a TPM role. For folks who are coming from non-technical backgrounds, held non-technical roles in the tech industry, or haven't utilized their tech education in their past experience, know that you can still build a TPM career. Aspiring TPMs can strengthen their technical expertise by putting in a lot of hard work to build the "T."

How technical does a TPM need to be?

This question often comes up and the answer is "it depends." While TPMs are not required to code, having an understanding of the system will make you a stronger TPM. You can be a TPM with a big "T" or a little "t" depending on the domain you are working in. For highly technically complex programs such as AI and ML, deep expertise will make you more effective. More experienced or generalist TPMs may not always need to flex the big "T" because they may bring a different core strength to the team. However, note that if you are a TPM working closely with engineering and want to have an influence, develop your technical prowess.

Programming is a skill best acquired by practice and example rather than from books.

— Alan Turing

Here are three ways that you can develop your technical domain expertise.

Take Courses: There are many free online sources like Coursera and Udemy where you can learn the basics about new technical areas like programming, databases, cloud architecture, etc. Check out Programming Interviews Exposed (Mongan, Giguère, and Kindler 2013) and Designing Data Intensive Applications (Kleppmann 2017). If you are in the software industry, take a programming course in Python to understand the fundamentals, or learn about

SQL. You can also take specialized courses on Generative AI and ML such as the ones produced by Coursera Chairman and Co-Founder Andrew Ng. Courses can be good for building conceptual knowledge without having to learn programming. If you intend to stay in your current field, get advice on which courses are best from your technical peers.

Leverage Technical Peers: Sometimes, learning can happen best by connecting 1:1 with a peer directly. Engineers are often excited to share their technical knowledge with someone who is willing to learn. Ask them to explain and or draw the system workflow or architecture. Ask questions if you don't understand something and then draw out the design yourself and see if you can explain it to someone else. Talk to multiple folks on the team - engineers, architects, and other TPMs - so you get a different perspective from each.

Learn System Design: Systems design is the process of defining the architecture, product design, modules, interfaces, and data for a system to satisfy specified requirements. Systems design can be seen as the application of systems theory to product development ("Systems design" Wikipedia). System design is one of the most important and feared aspects of software engineering. One of the main reasons was that everybody seems to have a different approach; there are no clear, step by step guidelines (Chakraborty 2020). You can also take a basic system design video course (Educative 2019) to dive deep.

It is important to thoroughly understand end-to-end system design as it is a key part of TPM interviews. Furthermore, as you manage an engineering program, understanding how the system works will help you identify dependencies, estimate effort, and anticipate key risks.

Here are a couple of ways you can strengthen your technical judgment on designing good systems.

- Dig into the systems in your existing company even if you are not in a technical role. Starting in an area with which you have some familiarity will help you connect the dots. Even though I come from an engineering background, I utilized this tip whenever I interviewed for a new role.

Example: If you are a business analyst in a financial institution, learn how the different components work together to form the end-to-end system to enable thousands of parallel transactions. Or, if you are a design program manager in Ads technology, explore more about the Ads ecosystem. Where does your company fall into it? How does ad bidding happen?

- Request engineers from your team to invite you to sit and listen in on technical design reviews. You will get a chance to hear the senior engineers' technical conversations, and if you are already familiar with the program, it will help you get new insights into the technical aspects. It will help you internalize the concepts and then leverage them if you take other courses.

- Practice system design problems that are outside of your domain. Check out 10 system design interview questions (Haq 2017) that you can practice building your muscle. Generally with any system design, you make sure you have thought through all the different use cases, edge cases, privacy, security, data flow etc. Practice with a colleague or friend to get comfortable in breaking down any design question and building a solution that is robust. Formulate your answers in a structured manner and show your thought process as most system design questions are open-ended.

PRO TIP: Keep up with technology trends even if you are an experienced TPM. This will allow you to target opportunities in different domains and industries.

MOVING FROM ENGINEERING

The TPM role can be a natural progression for engineers who want to be involved in the product development cycle from concept to launch. While the technical skill set can come easily, it is crucial for engineers to understand the driving forces behind pursuing a TPM position. A significant change engineers need to make when pivoting to a TPM role is shifting their mindset from an engineer's technical viewpoint to that of a TPM's program-centric outlook. It is also essential to recognize that TPM roles rarely offer the same compensation level as engineering positions, at least for now. Therefore, individuals considering this transition must have a genuine passion for the challenging nature of a TPM role, even if it may not provide the same career advancement opportunities or familiar role expectations as engineering.

> *Even in such technical lines as engineering, about 15% of one's financial success is due to one's technical knowledge and about 85% is due to skill in human engineering, to personality and the ability to lead people.*

> — Dale Carnegie

I know many TPMs who moved from engineering management to TPM and did not enjoy it because of the lack of control over outcomes. EMs have more direct authority over the engineering team but it's not the same for TPMs. Think about how you will continue to maintain and build relationships with engineers so that you can still influence them. Personally, I enjoyed being able to see the big picture and have a wider influence and impact on the business. So you may need to reframe your thinking from control and outcomes on specific teams to impact on multiple teams.

As you start thinking about moving from engineering to a TPM role, here are five steps you should follow:

Think like a TPM: Once you have decided to move into the TPM role, you will need to start thinking like one. That means you will need to step outside of your engineering mindset and expand your view of what is important in a program. As an engineer, you may think about the technical details, however as a TPM, you need to think beyond the technical implementation and dive into program complexity, dependencies, timelines, etc. If you are assigned a feature, ask questions about other features and how your feature is connected, or ask about what other dependencies exist. Even your resume can get hyper focused on technical details, so thinking like a TPM will help you get in the habit of TPM-relevant actions and impact.

Treat your career pivot as a program: Programs don't just have to be about products and services. If you are making a big change like this, start by building out a schedule with a breakdown of work and milestones to get there. For example, if you are migrating or deprecating an old legacy system to a new technology, what steps are required and how do you break it down? Apply that similar framework to migrate your old career to a new one. This will give you practice to plan out a program and execute it. Even if you made a pivot previously without a plan, consider these specific actions:

- Decide your timeframe for making the official move and work backwards from it.
- Do your research on what is needed and lay out the steps to get there.
- Talk to your manager about your goals and what support you need from them.

Start building your program management skills: Look for opportunities in your current role to take on bigger responsibility in programs. If you are a technical lead, you may already be doing some of the collaboration, estimation, and scheduling work with your team. Expand on your responsibilities. These programs can be within your domain, so you can still leverage your technical expertise while building new skills.

- Reach out to any existing TPM and ask them if you can help in any area. They will be glad to have more support.
- Talk to your manager about dedicating a percentage of your time to build the muscle for program management.
- Take program management courses or certifications to understand the fundamentals and how to apply them in real-world scenarios.

Hone your leadership and people skills: I have mentioned previously that soft skills are often the most important skills for being a successful TPM. As an engineer, you may not get a lot of opportunities to flex those muscles. Start identifying the areas that you would like to improve and seek opportunities to develop these skills.

- Build relationships with key stakeholders outside of your work meetings. This will help you understand their incentives and challenges.
- Strengthen up your presentation and public speaking skills if you are not used to doing so in your current role as an engineer.
- Refine and build your communication style and learn from observing successful people around you.

Find mentors: The best way to get to your goals faster is to find someone who can help you. They can be mentors who give you practical tips in the context of your team/company or someone well established in the TPM function. People are often more than willing to help. Having more than one mentor will not only provide you with different perspectives, but it will also give you more than one tool to handle tricky situations.

* * *

PRO TIP: Aim to find at least one mentor that is outside your company. You can do so by being part of TPM communities or attending networking events. This mentor can help you understand the variations in TPM roles in different companies, which will broaden your understanding.

CHAPTER 9

DIFFERENTIATING YOURSELF

*Six seconds is all you have to grab the attention of
a prospective recruiter or hiring manager.*

– Research Studies

Your journey now shifts to second gear, where your newly acquired skills and experience will be put to the test in job applications and interviews. You are likely to face stiff competition from individuals who are already in the field and have formal experience as a TPM. Even if you may not have multiple years of TPM experience, you have transferable skills and strengths that can be equally important. Assess your Unique Selling Point (USP) and differentiating factor - maybe it's a cutting edge tech domain, high-growth startup experience, or ability to work with senior leaders. Utilize your USP when you apply for TPM roles.

CRAFTING YOUR RESUME

As a hiring manager for technical program managers, I've encountered numerous poorly written resumes filled with errors, typos, and irrelevant details, which, unfortunately, create a negative impression. Reviewing hundreds of resumes also meant that if a resume doesn't resonate in the first half page, it got rejected. This is after passing through automated systems and recruiter checks.

Regrettably, even exceptionally qualified candidates often miss out on opportunities due to lackluster resumes that fail to effectively showcase their relevant experience.

When aiming to transition into the role, it is vital to emphasize your previous TPM-like experience and present your desire to take on the role along with your unique expertise. Think of your resume as the first step to make a strong impression and demonstrate why you are the perfect fit for the position. The resume needs to convey a compelling story about your skills, experience, and potential. Therefore, invest the time to create a well-crafted, concise, and error-free resume that truly reflects your capabilities and sets the stage for a successful interview.

Once you have that, tailor your resume to the job description so you can make sure that the most relevant requirements are showing up in your resume. Today, there are many AI-driven services that can help you with this so it is not as time consuming. At the very least, ask ChatGPT - the generative AI tool from OpenAI - to review your resume based on a job description and give you suggestions. Then apply your intuition to make sure it comes across as human.

If you are not officially in the TPM role, then make sure to adopt the mindset of a TPM and clearly articulate how your previous roles have prepared you for this position. Highlight specific achievements, programs, and responsibilities that align with the demands of a technical program manager. Emphasize your ability to navigate complex technical programs, manage stakeholders, and deliver results effectively. Furthermore, use your resume to communicate your genuine passion for the TPM role and your eagerness to contribute to the organization's success. Showcase how your unique expertise can add value and contribute to the growth of the team and the company as a whole.

Remember, your resume is a powerful tool that can open doors to exciting career opportunities. Make sure it speaks volumes about your capabilities and potential, leaving a lasting positive impression on potential employers. A well-crafted resume can significantly increase your chances of landing interviews and ultimately securing the TPM role you desire.

Here are some of the most common mistakes I have seen and recommend you avoid.

Formatting and Language

While most of these are found in all types of resumes, TPMs should pay extra attention as the resume also showcases your written communication skills, which is an important aspect of the TPM role.

Inconsistent formatting can make your resume appear cluttered and distract from the main content of your resume.

Example: Overly bolded phrases, multiple font styles, colors, images, and tables.

Led a **software development project** (digital twin model simulation) of aircraft from **initiation through closure using Agile methodology**

Resume Example with Unnecessary Bolded Phrases

Inconsistent language like poorly constructed sentences, inconsistent grammar, and incorrect usage of verb tense, nouns, or numbers.

Example: First two sentences in past tense followed by last two in different tenses. The last two sentences are listing only responsibilities and do not make the resume strong.

- Executed migration of user information for 10K users, from applications to a centralized database to bring program into green for security compliance.
- Refined release deployment process to reduce time spent unemployment by 20%.
- Develop user stories for upcoming features and business requirements.
- Planning and managing monthly releases.

Resume Example with Inconsistent Grammar

Technical jargon and industry specific acronyms can make your resume difficult to read and understand.

Example: *Usage of the acronym EIAM is not helpful and raises questions, especially for someone who may not be familiar with the specific industry.*

Delivered an EIAM automation Value Ad which actually saved 96% of the time in provisioning.

Resume Example with Technical Jargon

Lengthy and verbose resumes indicate a lack of clear and crisp communication. Resumes should be concise, ideally one to two pages, depending on your level of experience. A general rule of thumb is to limit your resume to one page if you have less than 10 years of experience. If you have more experience, the length can go up to two pages. However, focus on the last 5-7 years rather than including all accomplishments for every job. Avoid reducing your font size too much just to fit everything in.

Example: *If your first job was 15 years ago, just mention the role/title, company name and year. If your second job was 10 years ago, add 1-2 sentences describing the overview of the role and domain without listing out all programs.*

Program Manager/Technical Lead/Software Engineer, Blackbaud Inc.
- Managed team/projects for successful release and continued support of education and financial solution systems across desktop, web and cloud based applications.

Other experiences
• Product/Project Manager, Real Estate Startup	2010-2011
• Software Development Intern, Honeywell, Inc.	2002-2003
• Vice President of Publishing, Corporate Leaders Program, Arizona State University.	2002
• Intern, Indian Space Research Organization	2000

Resume Example Listing Early Experience in 1–2 Lines

Core Content

TPMs work on many different aspects of the program, so it can be tempting to include all the information to showcase your impact. However, that often can make the resume lengthy and take away from your core skills. It is also important to avoid the temptation to include every program or accomplishment in the resume. Instead, focus on your top three achievements per job that are most relevant to the role you are applying for.

Lacking a clear objective and your unique selling point (USP) at the top of your resume can hinder the reader in understanding your desired role and why you are the right person for it. TPMs often have a variety of experiences that can apply to different roles, so it is important to highlight your desired role and what you can do in that position based on your expertise.

Example: "*Manage innovative product development as a TPM by applying my diverse product and program management experience along with strong leadership, communication and technical skills.*"

Irrelevant content and too much detail about the program can take away from showing your core contribution. Not all of your experiences will be relevant to the role you are applying for, so it's best to focus on specifics that relate to the job. You can check the job description to assess what will fit best.

Example: You may not need to include the technical architecture details for a TPM job. Adding irrelevant details about activities performed that are not a core part of the role like "interviewed many candidates" or "participated in recruiting events" will not set you apart from other candidates.

Lacking a clear distinction between accomplishments and day-to-day responsibilities. TPMs are responsible for many program-related actions, however, focusing on the activities performed rather than the impact achieved does not help you stand out from other applicants.

Example: Phrases like "Conducted standups," "Created Jira tasks," "Held weekly meetings," or "Wrote status reports" do not show how those actions led to results.

Weak content can trivialize your accomplishments and make you appear less capable.

Example: Using weak verbs such as "coordinated," "facilitated," "helped," or "supported" instead of powerful verbs like "led," "drove," or "spearheaded."

Using too many subjective adjectives and adverbs like "several," "many," "significantly," "faster," "much," or "extremely" instead of hard quantifiable data and objective language like "10X improvement," "collaborated with 25 engineers," or "reduced cost by 3X or 25 million."

Treat your resume as the foundation to your career ladder. If the first step is broken, how will you get to the next step?

PRO TIP: Always share a PDF version of your resume, so it is compatible across operating systems and devices. You don't want your hard work to get messed up just because someone doesn't have the software to open a Microsoft Word file.

STRATEGIZING YOUR JOB SEARCH

As I was doing research for this book, I found all the job application emails from the time I was actively searching for a new role. I was applying for many different types of roles - product manager, program manager, project manager, and business analyst with the same resume. I was also completely unaware if the level and requirements matched my experience. I found so many emails where I had applied for Principal Program Manager roles at Microsoft and some other companies. No wonder it took me so long to secure a job! I was not intentional about understanding what could increase my chances of success.

Job searches can be a daunting process that requires a significant investment of time and effort. As I mentioned earlier, transitioning from an engineering role to TPM took a long time for me. Looking back, I realize I did not have a sound strategy. I just knew two things - I wanted to find a program manager role that was technical and I wanted to move to a certain area of the country. I encourage you to develop a clear understanding of your narrative, strengths, and requirements, so you target the right jobs and increase your chances of success. Here's what I recommend:

Perfecting Your Pitch

Begin by defining your goal: What you want to do and where you want to go? Starting from the destination can help showcase your intent and what you need to get there. Use the following structure to clarify your objectives:

- **Industry:** Which industries do you want to work in? Where can your experience make the most impact?

- **Company/Mission:** Which are your dream companies? What kinds of companies align with your values and mission?

- **Domain:** Which areas or technologies are you interested in exploring? How does that relate to your current experience?

- **Role/Title:** What is the role and title you seek? What are the key responsibilities you want to have?

- **Level:** Which level are you targeting, and how does it align with your current level? What do you need to get to your desired level?

- **Compensation:** What is the minimum compensation you need, and what is the ideal compensation that would make you give up something else on the list above?

Once you have answered these questions, prioritize them from most important to least important. This ranking will help you save time by focusing on job applications that align with your priorities. If a job description does not meet your criteria, you can skip it and move on to the next. This will also save you time and energy, which can be used to focus on increasing your odds elsewhere.

Managing Career Pivots

Changing your career path can be both thrilling and daunting at the same time. There are three key factors to consider when making a career pivot: your role, the company you work for, and the field or domain you are in. To make the process smoother, it's a good idea to only change one or, at most, two of these factors at a time. This will help you avoid feeling overwhelmed and boost your chances of success.

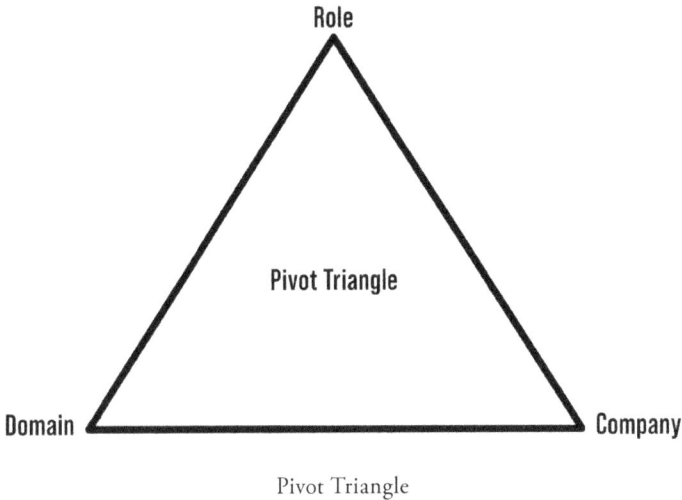

Pivot Triangle

Role: If you are transitioning to a TPM role, try to leverage your current domain expertise at your existing company. You should still focus on aligning your current skills and experiences with the core aspects of the TPM role and highlight how your previous roles have prepared you for these challenges.

Domain: If you are looking for a TPM role at a different company, take advantage of your existing domain knowledge. Your domain knowledge will also help increase the probability of acing the interviews. Showcase how your in-depth understanding of the domain enables you to hit the ground running and contribute to the success of ongoing and upcoming programs.

Company: If you are changing both role and domain, try to do so within the same company as you already have an understanding of the company's culture, products and internal processes. Look for internal TPM opportunities where your background knowledge can add immediate value.

The pivot triangle concept can apply to any career transitions irrespective of which role you are targeting.

Leveraging LinkedIn

I found my TPM job via LinkedIn. I was on the platform all day during that initial transition. I also wanted to move across the country, so I changed the location on my LinkedIn profile to that area. That small change is what got my foot in the door because recruiters generally add a location filter to their searches. Of course, I was honest with them during the process. Had I not been strategic, I would not have got that first call. Of course, today, with remote work becoming more common, the location aspect may not matter as much. Irrespective of the location factor, there is no argument that LinkedIn is an indispensable tool for enhancing your job search and achieving successful career pivots, especially when seeking a TPM job.

To maximize your chances of finding the right opportunity, it's crucial to strategically position yourself on the platform and actively engage with your network.

Optimize your profile: Your LinkedIn profile is your professional online resume, so ensure it is up-to-date with relevant information that highlights your skills and experiences. Use a professional profile picture that conveys a positive and approachable image. Craft a clear and impactful headline that accurately reflects your current position or clearly indicates the type of TPM role you are seeking. Your headline acts as a virtual handshake, making it essential to create a strong first impression.

Write a compelling summary: In the summary section, succinctly showcase your key achievements, skills, and specific experiences that align with the TPM role. Use compelling language to capture the attention of recruiters and hiring managers. Consider including specific examples of successful programs you've managed (e.g.: launch X program spanning Y teams), stakeholders you've collaborated with, and the impact you've made in previous roles. This section is your opportunity to tell your professional story and demonstrate why you're an ideal candidate for a TPM position.

Engage and expand your network: LinkedIn is not just a static resume platform; it's a dynamic professional community. Engaging with your network and participating in industry discussions can significantly boost your visibility and reach. Join relevant groups related to TPM, program management, and your specific domain to connect with like-minded professionals and industry leaders. Engaging in meaningful conversations shows your passion for your field and keeps you informed about the latest trends and developments.

Showcase expertise: Establish yourself as an expert in your industry by sharing articles, insights, and updates related to TPM and your domain. Providing valuable content showcases your depth and demonstrates your commitment to staying informed and relevant in your field. It can also attract potential employers who recognize your knowledge and potential contributions to their organizations.

Network with intention: Actively reach out to people in your network or those working in companies you aspire to join. Attend virtual events, webinars, and conferences where you can network with professionals and recruiters in the TPM space. Be genuine in your interactions and consider requesting informational interviews to gain insights and advice from TPM professionals who have successfully navigated similar career transitions.

Seek recommendations: Request recommendations from colleagues, supervisors, or mentors who can vouch for your skills and work ethic as a TPM. Positive endorsements add credibility to your profile and enhance your chances of being noticed by recruiters.

* * *

By thoughtfully positioning yourself on LinkedIn and actively engaging with your network, you increase your visibility, build your personal brand, and attract relevant job opportunities. An effective

LinkedIn presence can establish you as a valuable asset in the TPM field and open doors to exciting career possibilities.

PRO TIP: Mark yourself as "Open to Work" either publicly or privately so that you show up higher in recruiter searches.

CHAPTER 10

CRACKING TPM INTERVIEWS

One important key to success is self-confidence. An
important key to self-confidence is preparation.

– Arthur Ashe

Having been on both sides of the interview table, this book would not be complete without sharing with you my experience and expertise on nailing your interviews. In the early days of TPM, there weren't a lot of resources about the role, let alone about how to interview for them. The TPM interviewing frameworks were somewhat haphazard, partially adopted from engineering, partially from the Microsoft style of interviews - *explain why manhole covers are round and design a teacup.* Once again, Amazon took the lead in creating a good framework for TPM interviewing. Most companies, big and small, have adopted a somewhat similar structure. The specific interview types may have different names but all are trying to assess the core skills of a TPM. As a candidate, I have given hundreds of interviews and I have also interviewed hundreds of candidates. In my previous roles, I have been involved in creating interviewing frameworks and refining them to ensure we have a clear and objective way of identifying and hiring the best candidates. I have also trained many TPMs on interviewing skills, enabling them to land their dream jobs. As you read through this chapter, keep in mind that the content covered in the rest of the book is equally useful to crack those TPM interviews. I will give you an overview of the TPM interview process but will not go into every detail or provide a question bank. You can find those via my website.

As you know, TPMs work on complex, cross-functional programs and are responsible for driving product development from conception to launch. As such, the interview process for TPM positions is rigorous and typically consists of multiple rounds of interviews with various stakeholders involving a range of technical and non-technical assessments. TPM interview questions cover a wide range of topics, including program management, technical expertise, system architecture design, stakeholder management, communication, and leadership behaviors. TPM interviews aim to evaluate a candidate's ability to manage complex programs, communicate effectively with cross-functional teams, think strategically about business objectives, and lead and influence teams. I remember Google used to ask algorithm- and data structure-related questions to test technical knowledge. Smaller companies often ask candidates to create a presentation about a fake program, so they can assess candidates for multiple skills.

INTERVIEW TYPES

Most companies have 5-6 interviews after the typical recruiter or hiring manager phone screen.

These interviews are categorized as the following:

- **Technical Aptitude/Retrospective:** Tests the ability to provide specific examples and understanding of the technical domain from a previous experience. Interviewers are seeking to understand your ability to steer technical decisions to best meet product/technical requirements. This interview is often focused on your past experience or a familiar domain.

- **Technical Judgment/System Design:** Aims to test your understanding of system architecture and ability to articulate design choices and tradeoffs in an unfamiliar technical domain, or most likely the domain in which the company operates.

- **Program Sense/Management:** Tests your foundational program management knowledge and methodologies. Interviewers will touch upon all phases of a program management lifecycle from planning to execution to launch. The job responsibilities outlined are a good indicator of the concepts that will be part of the questions.

- **Partnership/Stakeholder Management:** aims to test your ability to build relationships, create a broad consensus on contentious issues, manage conflicts, and negotiate effectively to create win-win solutions.

- **Leadership:** Encompasses multiple skills such as critical thinking, analyzing, and arriving at conclusions independently, making decisions, influencing at all levels, and having the ability to deal with ambiguity to solve complex problems.

- **Behavioral/Values/Bar Raiser:** Aims to seek individuals that align with the company values, mission, and vision. Interviewers want to understand if you are someone who would make the company better and someone who they would like to work with.

PRO TIP: For details about each of these interview types and how to prepare for them, check out the "Cracking the TPM Interview" course available on my website.

INTERVIEWING BASICS

Mastering technical program management interviews requires a methodical approach and thorough preparation. Yet, I find that many candidates rush to give interviews and often miss the obvious things about nailing interviews. Here are nine strategies you can apply to increase your chances of success in any type of interview.

Research the company: Show your genuine interest in the company by thoroughly researching its mission, values, and recent accomplishments. Understand how your expertise aligns with the organization's goals and be prepared to articulate how you can contribute to its success. Tailor your responses to showcase your understanding of the company's challenges and how your skills can address them.

Give yourself ample preparation time: Technical program management interviews demand substantial effort and practice. Start your preparation well in advance, allowing yourself enough time to review your experiences, skills, and relevant projects. Take advantage of resources like online guides, interview books, and mock interview sessions to fine-tune your responses. If you are not ready, try to push out scheduling any interview. It is better to push out rather than get rejected as most companies often have a freeze period of 6-12 months for candidates who did not pass.

Practice your answers: Practice is the key to building confidence in interviews. Without practice, it is easy to veer off course, ramble, and come across as unprepared. Practice telling your examples of past experiences, challenges faced, and successful outcomes. You can do this using video, a mirror, or a friend. Get feedback from them and improve on your answers. Be sure to emphasize your problem-solving abilities, collaboration skills, and leadership qualities in these examples. The more you rehearse, the more comfortable and articulate you will become when sharing them during the actual interview. You can even do mock interviews with other TPMs in your network or with a professional. I often enlist my partner to practice with me. In fact, we would go for long walks and I would practice answering questions put forth to me. This preparation helped immensely in presenting my examples in a clear and succinct manner. I want to add here that while you want to prepare and practice; you don't want to memorize the responses and come across as mechanical. Practice being natural as well.

Demonstrate thoughtful and curious thinking: During the interview, take your time to think through questions carefully. Technical program managers need to exhibit thoughtful decision-making and analytical skills. Show your curiosity by asking follow-up questions to gain a deeper understanding of the problems or scenarios presented. I even recommend taking a pause or asking for 15-30 seconds before you answer. This way, you can put a thoughtful answer together that resonates with the interviewer.

Be specific and data driven: Be as specific in your examples as possible. Support your responses with concrete data and metrics. Discuss the quantifiable impact you've made in previous roles, such as program outcomes, cost savings, process improvements, or team performance. Data-driven responses highlight your ability to use evidence to inform decisions and measure success.

Provide examples of results achieved: TPM roles are results-oriented, so emphasize your ability to drive programs to successful completion. At a minimum, use the STAR method (Situation, Task, Action, Result) to structure your answers when discussing your accomplishments. Highlight the challenges you faced, the actions you took, and the measurable results you achieved.

Ask insightful questions: The interview is not just an opportunity for the company to assess you; it's also your chance to evaluate if the role and company are the right fit for you. Prepare thoughtful questions about the day-to-day responsibilities, team dynamics, and potential growth opportunities. Engaging in meaningful dialogue demonstrates your genuine interest and commitment to making an informed decision.

Example: Don't just ask what a day in the life of TPM looks like. Instead, ask "How does the team make decisions and what role does a TPM play in those decisions?"

Express your enthusiasm: Throughout the interview, let your passion for the company and technical program management shine through. Enthusiasm for the role and the company can leave a lasting, positive impression on the interviewers. Show them that you are genuinely excited about the prospect of contributing to their organization as a TPM.

Perfect your storytelling: Practicing your answers is required, but you want to ensure that those answers are structured in a way that makes a lasting impression. Your answer contains more than just examples of actions and activities - they are stories - and stories, when told correctly, can be very powerful. In the next chapter, I go into more detail on why STAR format is not enough to tell compelling stories.

PRO TIP: Incorporate company values into your examples to show how you can fit with the company culture. This is especially important for the behavioral or culture fit interviews.

REFLECTING ON YOUR INTERVIEWS

It took me more than a year to land my first TPM role. Fast forward a few years later, it took 50+ interviews to get into my dream company. That means, I didn't pass a number of interviews and with every such interview; I retrospected on what didn't work and learned something new. I applied those learnings to the next interview. Having been on the other side of the table, where I have had to reject many candidates, I can tell you that there can be many reasons for not getting an offer and sometimes, those reasons may have nothing to do with you.

I know job interviews can be incredibly intimidating and stressful, often causing fear of failure, judgment, and forgetfulness similar to public speaking. The emotional, financial, and professional stakes can be high, which only adds to the pressure of performing well. The truth is that acceptance rates are often less than 5%, so it's important to remember that not passing an interview is not defining you or your character. Every such interview is an opportunity for growth and development. In fact, I believe that there's always something better around the corner.

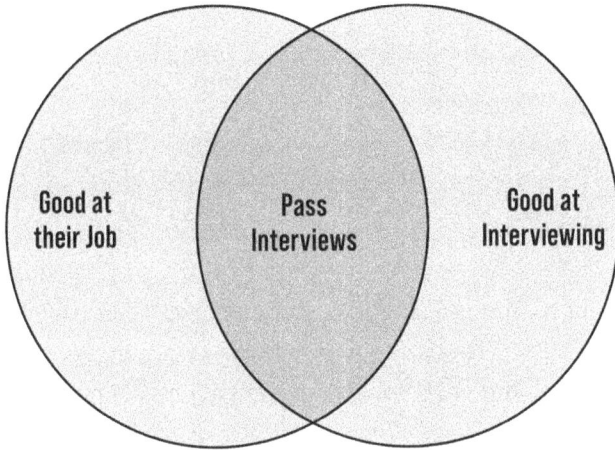

Interview Success Zone

Interviewing is not just about how skilled you are at your job; it is also about how effectively you market yourself to potential employers.

Reflecting on interviews that did not go well can illuminate new paths and ideas for you. It will give you the energy to push ahead. Consider these five factors if an interview doesn't result in the outcomes you were hoping for:

Your desired position does not leverage your current expertise. When you are perceived as an expert, people are more likely to want to bring you in. If your experience or expertise is not relevant to the position, you may not come across as a strong candidate.

Your level expectations are not in check. Even if you aspire to get at least one level up when joining a new company, that doesn't always mean you are ready for that level from the company's perspective. If you're making an industry switch, be conservative in level targeting, as higher levels come with higher expectations and performance pressure. Instead, you can focus on growing in other ways and target a higher level once you're familiar with the culture and expectations of the new company. Remember that levels are not uniform across companies and are a function of scope, complexity, and impact.

$$f \text{ (Level)} = \{ \text{ Scope + Complexity + Impact } \}$$

Team or company culture may not be a good fit. It's essential to assess your behavioral and culture fit during interviews. While technical or management skills can be taught, a lack of cultural fit can be challenging to change. Companies are looking for individuals who can adapt and work for their mission with their stated values, but cultural fit is not only about you; it's also about the company's way of operating. Make sure to "interview" the company's culture as well and make your decision accordingly.

Your skill level is not aligned with the job requirements. Identify areas of improvement including technical, communication, leadership, and soft skills, and work on them. Building self-awareness of your blind spots will be valuable in growing yourself. You can get feedback on your mock interviews with friends, family, trusted colleagues, or a professional to help you improve your skills. Building your skill set in your current job is another way to gain real experience.

You did not give yourself the gift of time. Time is crucial when preparing for interviews. If you're looking for a new job while still working, you may be struggling to find the time to improve your skills or prepare adequately for interviews. But, it's essential to carve out time for yourself to enhance your skills, work on your resume, practice mock interviews, and research potential companies. Finding a balance between your current job and your job search can be challenging, but giving yourself enough time will make a difference in your job search success.

* * *

PRO TIP: Don't interview at your dream company first. The stakes are too high, so instead interview at other companies as a way to get practice and get over your interview jitters.

CHAPTER 11

POWER²FUL STORYTELLING
FOR SUCCESS

Story, as it turns out, was crucial to our evolution -
more so than opposable thumbs.
Opposable thumbs let us hang on;
story told us what to hang on to.

– Lisa Cron

When I started interviewing candidates, I observed the nuances of their responses that helped me better connect with their stories and assess if they were the right candidates. I found that the best examples were the ones that were highly engaging with a compelling narrative that showcased how the candidate overcame challenges and used their skills to achieve results.

Storytelling is not just for keynote speakers, brand marketers and influencers. Everyone needs to learn the art of storytelling because it is an indispensable tool. When it comes to captivating interviewers and leaving a lasting impression, storytelling can be the difference between getting an offer versus a rejection. Interview storytelling entails crafting narratives that not only capture the attention of the interviewer but also vividly convey your unique value and achievements. While the STAR method serves as a popular framework for answering interview questions, I believe that relying solely on it may fall short in today's modern dynamic workplace.

In my experience, the simplicity of STAR often fails to encapsulate the intricacies of complex situations, potentially reducing your stories to mere chronological accounts of events, which can disengage the interviewer. To truly make an impact during interviews, it is essential to transcend the confines of the STAR method and embrace a more holistic approach to storytelling. An effective interview story extends beyond a mere sequence of actions; it goes deep into the emotions, challenges, and transformational journeys you experienced along the way. Good stories, in this context, possess several key qualities: they captivate the listener's imagination, evoke empathy, illustrate your problem-solving abilities, highlight your adaptability, and showcase the tangible results of your efforts.

I created the POWER^{2}ful Storytelling framework to help candidates build their interview storytelling skills. This framework has proven to be so powerful (pun intended) and a game changer for hundreds of candidates. To create POWER^{2}ful interview stories, consider the following elements:

- **Compelling context:** Set the stage by providing a concise yet compelling overview of the situation you faced. Capture the interviewer's attention by painting a vivid picture of the challenges, the stakeholders involved, and the overarching goals. Strike the right balance in the level of detail and amount of technical jargon, keeping in mind your audience.

- **Personal impact:** Emphasize the personal investment you had in the situation. Describe how you felt, the level of responsibility you undertook, and the motivation that fueled your actions.

- **Conflict and resolution:** Highlight the hurdles you encountered and the strategies you employed to overcome them. Focus not only on the actions you took but also on the creative problem-solving, critical thinking, and resilience you demonstrated.

- **Collaborative approach:** Showcase your ability to work effectively with others by illustrating instances where you leveraged teamwork, communication, and influence to achieve shared objectives. Highlight the successful partnerships and collective accomplishments you were part of.

- **Measurable impact:** Clearly articulate the quantitative and qualitative outcomes of your efforts. Highlight how your actions generated tangible results, such as increased efficiency, cost savings, improved customer satisfaction, or successful program completion.

Let's use an example to demonstrate the difference:

Question: Tell me about the most challenging partner you have worked with in your career.

STAR	POWER²ful STORYTELLING
I was working on the launch of an AI program alongside a product manager. The product manager was new to the program while I had been the program TPM for three months. The PM started duplicating my efforts or interjected my communication to leadership.	*I was working on a very challenging AI program that had a tight deadline. We had a new product manager join the program midway. They were new to the company, so while they needed to ramp up, they also needed to make this program successful.*
There was confusion about the role and responsibility. So I went to my manager and escalated the issue. The manager talked to their manager and resolved the situation.	*Since the program had very high visibility, our senior leaders had to be kept in the loop. I was working on multiple interconnected projects that required constant communication. I started noticing that the PM would interject my communication or respond in my stead. I realized that there might be confusion about the role and responsibility. So I scheduled a 1:1 to*
As a result, I was able to continue with my work and launch the program on time.	

STAR	POWER²ful STORYTELLING
	talk to them and asked them about some of their actions to get clarity. However, they got very defensive and the conversation did not go well.
	The PM kept on repeating their behavior, which was essentially undermining my work. I tried to have another conversation to lay out what each of us could do that would still be impactful while avoiding stepping on each other's toes. I mentioned that our goal was common and there was enough work for both of us. However, there wasn't a resolution and since the launch was a week away, I focused on ensuring its success. Some of the behavior was also noticed by our leaders, so once the launch was done, I connected with our managers and discussed the events.
	After a few days, I once again went to them to have a conversation about how we want to move forward while acknowledging the pressure they must have been under as a new PM. We talked and resolved the issue eventually, but it was a very challenging time period.

STAR vs. POWER²ful Stories

As you can see, STAR seems succinct but doesn't really showcase the challenge and conflict in the same way. The personal impact is not apparent and seems like it got resolved quickly and easily with escalation. However, the POWER²ful format gives more context and talks through the struggle of resolving the conflict directly and not resorting to escalation at the get-go.

* * *

By incorporating these storytelling elements into your interview responses, you can go beyond the traditional STAR method and create a narrative that resonates deeply with the interviewer. Remember, it's not just about sharing what you did, but also conveying the essence of who you are as a professional and how you can contribute to the prospective role and organization. You can even leverage this framework for situations beyond interviewing, like personal branding, presentations, etc.

PRO TIP: Check out the complete version of the POWER²ful Storytelling framework available on my website.

CHAPTER 12

MASTERING OFFER NEGOTIATION

Everything is negotiable. Whether or not the negotiation is easy is another thing.

– Carrie Fisher

For my very first job right out of graduate school, I got an offer for an entry level software engineer role. I knew a friend of mine had negotiated 10% more in the same company for the same role. But I knew I was going to take the job. It didn't even occur to me that I could negotiate. I remember the hiring manager calling me to ask if I had any questions before I signed the offer letter. When I told him I just signed and emailed it, I could hear his voice - a little bit of surprise, a little bit of relief, and probably a little bit of pity. At that moment, I felt he was ready for a negotiation and maybe even raise the offer but I totally blew it. In the coming years, I always negotiated, but I still felt uneasy. I think as a woman, it is harder to negotiate. I often felt that if I asked for more money, they would have higher expectations of me on the job.

Negotiation is a nerve-wracking experience and most people don't like it. Most people are not born with negotiation skills either. Negotiation where financial gain is concerned is a totally different ball game than negotiating for resources as a TPM. In one of my MBA classes about negotiation, we were told that you can negotiate anything in a job offer - even the number of vacation days. Now, I haven't personally tried that one and it may not even work today. However, what I want to emphasize is that we often don't know what the other side is willing to give up. It can often feel like we are at the "desperate" end - after all, we are the candidates who need a job. But remember, depending on your interview performance, expertise and skill set, the company may need you badly as well.

I encourage you to get past the discomfort of negotiation. Trust in the value you bring to the table, your skills, your expertise and create a framework for yourself, so you know that whatever decision you make is good for you in the long term. Mastering the art of offer negotiation will ensure that you receive the best possible compensation and benefits package you deserve. Whether you're negotiating salary, benefits, or other perks, employing the following strategies can lead to a successful outcome.

Here are 11 tips to help you get prepared and get what you want and deserve. Most of these are applicable to anyone negotiating a job offer, however, as TPMs you can consider it practice to improve your influence and communication skills.

- **Research thoroughly:** Conduct in-depth research on the company's compensation and benefits packages, as well as industry and location benchmarks for similar positions. Utilize resources like Glassdoor and levels.fyi to gain valuable insights into typical earnings for your role.

- **Determine your walkaway number:** Prior to negotiations, establish your walkaway number, the minimum compensation or benefits you are willing to accept. Knowing this figure will help you remain focused and avoid settling for an offer that falls below your minimum acceptable level.

- **Avoid providing numbers first:** Whenever possible, let the other party present the first offer. This allows you to gauge their initial position and gives you more room to navigate the negotiation process. Ideally, you have a rough idea before going into the interview cycle itself. If you must provide a number first, offer a range to maintain flexibility and let the bottom of that range be higher than your walkaway number. Many states are now required to post salary ranges, which should help in setting expectations.

- **Show enthusiasm and interest:** During negotiations, express genuine excitement about the opportunity to join the company and contribute to its success. Demonstrating enthusiasm and interest can foster a positive negotiating environment and build rapport with the hiring team.

- **Identify must-haves versus nice-to-haves:** Clearly define your must-have items, such as a minimum salary or essential benefits, and distinguish them from nice-to-have items that are desirable but not deal-breakers. This prioritization will help you focus on crucial negotiation points and avoid compromising on essential needs.

- **Consider total compensation:** When evaluating an offer, don't solely focus on base salary. Take into account the entire compensation package, including bonuses, stock options, benefits, and potential for growth within the company. Think about how each component would benefit you in the long run.

- **Articulate your value:** Clearly communicate the value you bring to the company through your skills, experience, and achievements. Showcase how your contributions align with the company's objectives and justify the compensation you are seeking.

- **Practice assertiveness and flexibility:** Be assertive in advocating for your needs while maintaining a cooperative and respectful approach. Be prepared to compromise on certain points if it aligns with your overall career goals.

- **Get it in writing:** Once you reach an agreement, ensure that all negotiated terms are clearly outlined in writing. This helps prevent misunderstandings and provides a reference point should any issues arise in the future.

• **Express gratitude:** Regardless of the outcome, express gratitude for the opportunity to be considered for the role and for the time and effort invested in the negotiation process. A positive attitude can leave a lasting impression and foster goodwill.

TPM Compensation by Level

Your interview performance will determine the level at which you join the company, which in turn, will determine your compensation. Candidates have to demonstrate relevant experience in managing complex programs at the competency level required for the specific position. In most cases, companies do not negotiate the assigned level once an offer is made. Therefore, the best approach to secure the desired level is to hit the ball out of the park in all aspects of the interview process. You can still negotiate compensation within the range for a particular level.

The compensation structure for TPM roles varies across companies. The company size and maturity will determine the compensation bands at the different levels. You can use the high level breakdown of TPM compensation by level to help you with your offer negotiation tactics.

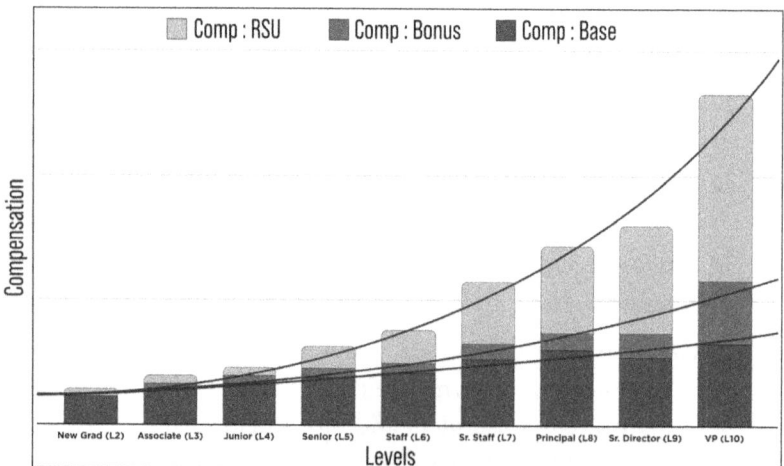

TPM Compensation by Level

The TPMs levels can be classified as broadband levels, such as Senior, Staff, and Principal, or numbered levels, like L1 to L10. Although the exact mapping of broadband to numbered levels may vary, a general approximation is shown in the figure. The compensation package for TPMs comprises a combination of base salary, bonus, and equity, with total compensation increasing as one progresses to higher levels. At more advanced levels, a larger portion of compensation is often provided in the form of equity.

* * *

PRO TIP: Be respectful of the recruiter and hiring manager's time. If you want to negotiate on multiple things, lay them all out at once. Do not draw out the process unnecessarily.

📣 TPM SPOTLIGHT

MALVIKA SINHA, SENIOR TPM

Malvika has been in the tech industry for over a decade, with her journey starting as a systems engineer. Over time, she transitioned into automation engineering and even served as a scrum master before finding her true passion as a TPM. She has spent eight years as a TPM, working in various industries including infrastructure, semiconductors, payments, compliance, healthcare, and gaming. Currently, she is in a TPM role at a renowned gaming company.

What motivated you to move to TPM?

Malvika: I thoroughly enjoy collaborating with people and discovering their unique stories, such as their backgrounds and motivations, as well as finding ways in which I can support them within my capacity. Additionally, I have a constant curiosity about the company's goals and the direction the leadership intends to take the organization. I love digging deep into metrics and knowing what would move the needle.

You mentioned you had a challenging journey to becoming a TPM. How did you overcome the struggles and failures during the interview process?

Malvika: Yes, becoming a TPM was indeed a challenging process. I had a strong background in engineering, particularly in infrastructure, but I lacked experience in building an entire system of architecture. To prepare for TPM interviews, I utilized my past program management experience and leveraged external resources like materials and mock interviews. I relied on my engineering friends for system design mock interviews. I also found many folks on community platforms like Blind, LinkedIn, and Discord who were eager to help out. The technical part of system design was the most challenging aspect for me, but I dedicated 3-4 months to daily mock interviews, took notes on feedback, and identified patterns to improve.

In the end, my hard work paid off, and I received offers from Facebook, Google, Stripe, and Airbnb.

What surprised you the most as you navigated the TPM role?

Malvika: The most surprising aspect was the diversity of the TPM role in different industries. There is no one-size-fits-all definition of a TPM's responsibilities. It can be more technical or more focused on business and end-to-end processes. It's crucial to work closely with leaders and stakeholders to understand their expectations. Building rapport with the people around you is essential for success in this role.

How did you create impact and get results in your role as a TPM?

Malvika: When I transitioned to managing larger teams with more engineers, I faced a lack of additional bandwidth, and there were no managers to support me. To address this, I built strong relationships with Engineering Managers (EMs) and aligned on RACI (Responsible, Accountable, Consulted, and Informed) frameworks. Instead of meeting with every engineer, I focused on collaborating with EMs to drive impact. I also piloted changes with one team before expanding them to others, and I adapted my methods based on the situation. Dealing with people's challenges head-on also helped me grow and develop as a TPM.

Another factor that has significantly contributed to my ability to create an impact is understanding which features or products will have the most impact on end customers and whether it aligns with the goals of other stakeholders. Establishing alignment with various stakeholders, including those in product, engineering, revenue, and cross-functional teams regarding priorities, has been of great value. While it may appear straightforward, there is often a disconnect, demanding comprehensive and timely communication, particularly for high-profile features.

What do you enjoy the most about being a TPM?

Malvika: I love the dynamic nature of the role. As a TPM, I'm always on my toes, aware of what's going on across various projects, and connecting the dots to ensure everything aligns with the big picture.

On the other hand, what aspects of the TPM role do you dislike?

Malvika: The role can be quite complex, and there's often not a lot of control over external factors. TPMs are accountable for various aspects like people, infrastructure, legal, and product, which can be challenging to manage at times.

How can organizations best leverage TPMs?

Malvika: TPMs should be partnered with leaders to fully understand the business objectives and anticipate potential challenges. It's crucial to give TPMs end-to-end responsibility and ownership over their programs. Additionally, organizations should balance the number of programs assigned to each TPM based on size and complexity to avoid spreading them too thin.

What advice would you give to other TPMs or aspiring TPMs?

Malvika: Don't be afraid to ask questions and communicate directly with engineering teams. Spend time with product managers to align your program goals with the overall product vision. Also, explore and use different tools, such as Smartsheets and Jira dashboards, to ensure accountability and success in your role. Finally, I know that since we are TPMs, we tend to dive right into problems to solve them! But I think it is good to take some time to understand the culture (which is very important), people, and processes before undertaking anything significant. It is okay to "Go slow to go far."

WON CHOE, TPM MANAGER

Won is currently a TPM Manager at one of the biggest online retail companies, responsible for leading a TPM function that works with a software team of 200 engineers. Prior to that, he worked at a multinational technology company, where he led a TPM team for region build projects, and also has experience with capacity planning.

What inspired you to transition into the TPM role?

Won: The motivation to become a TPM stemmed from doing tasks I genuinely enjoyed, some of which were TPM-related. I also wanted to dive deeper into the tech industry and learn more about technology. Embracing ambiguity, influencing without direct authority, and asking critical questions were skills I excelled at, making the TPM role an appealing fit for me.

Your journey to becoming a TPM sounds quite unique. Could you share more about it?

Won: My journey to TPM took an unexpected route. I initially studied to become a chemical engineer, but at that time a lot of the jobs for chemical engineers were at large oil companies (e.g. Chevron) that required working in remote places like offshore platforms. I decided against working at sea for extended periods and settled on working in environmental consulting as it was located near a major city. When the 2008 recession led to a layoff, it took me about a year to find a job as a business analyst at a small tech company, working closely with the CEO. Though the job wasn't directly tech-related, it gave me exposure to tech. I found the fast-paced environment and the focus on solving business challenges through technology fascinating. That's when I decided I wanted to make a career in the tech industry.

I gained experience working with tech-savvy individuals, which led me to exploring TPM roles at some of the biggest tech companies where I got the opportunity to work in different domains and eventually become a TPM manager. Throughout my career, I've relied on self-study and on-the-job learning, though I don't code.

As you built your expertise in TPM, what approaches did you take to develop your skills?

Won: My journey as a TPM involved various learning methods. I gained valuable insights through whiteboard discussions, engaging with engineering and tech teams, and building strong relationships with Engineering Managers. I would often request them to walk me through systems to acquire foundational knowledge. Additionally, I actively pursued self-study by enrolling in courses and watching educational content on platforms like YouTube. Implementing what I learned and continuously iterating based on practical experiences led to several "aha" moments in my TPM journey.

What surprised you the most as you navigated the TPM role?

Won: The most surprising aspect of the TPM role was its inherent ambiguity. Different teams utilize TPMs in various ways, and there's no fixed box that defines the role. This variability presents both challenges and opportunities, allowing TPMs to explore diverse programs and take on different responsibilities.

What aspects of being a TPM do you particularly enjoy?

Won: I see two types of TPM roles - a team TPM assigned to a specific team and focused on their programs only and an end-to-end TPM that drives programs across multiple teams. Personally, I enjoy working across multiple teams, having a breadth view, and full end-to-end ownership as it offers more variety in programs and responsibilities, making every day dynamic and engaging.

On the other hand, what challenges do you face as a TPM?

Won: One significant challenge in the TPM role is navigating career growth. Unlike other functions, TPM career paths may not be as well-defined, making it difficult to identify the next steps and career progression. This lack of clarity can be challenging for some individuals. Moreover, the requirement to influence without direct authority can be demanding and sometimes exhausting.

How can TPMs be leveraged most effectively within an organization?

Won: In my experience, TPMs are most effective when they are placed in large programs with multiple stakeholders and have the independence to operate outside specific teams. This impartiality allows TPMs to provide unbiased perspectives, ask critical questions, and ensure the success of complex programs.

How do you scale yourself as the lead TPM and how do you enable your team to do the same?

Won: Scaling myself as a TPM involved a few key strategies. With large, end-to-end program ownership, I made sure to provide my team with opportunities to dive deeper into specific domains, assigning TPMs to focus on particular areas of expertise. This balance between breadth and depth allows TPMs to develop a strong understanding of their domains while still having a broader view of the overall program. Given the inevitable time constraints, I focused on understanding key performance indicators (KPIs), metrics, in addition to the Objectives and Key Results (OKRs) to pinpoint potential issues. Utilizing subject matter experts and collaborating with others became essential in tackling complex challenges.

As a TPM Manager, what was different about the role compared to being an Individual Contributor (IC)?

Won: The transition to a TPM Manager involved a shift in focus. As an IC, my primary concern was my program and its output. However, as a people manager, success is defined more broadly. Success now means setting up the team for scaling, enabling them to overcome obstacles, and providing support beyond just executing tasks. I also consider factors such as the growth and well-being of my team as success metrics as a manager.

How do you support the growth and development of your team members?

Won: Most companies have effective mentorship programs in place, which I encourage my team members to participate in. Additionally, shadowing and a buddy system during onboarding prove beneficial for new team members. One challenge is that many TPM skills are intangible, making measurement difficult. Despite this, I focus on providing opportunities for growth, even if it means stretching team members slightly outside their comfort zones. It's okay if they don't fully succeed, as long as they learn from the experience.

What advice would you give to other TPMs or aspiring TPMs?

Won: My advice would be not to wait for the official title of TPM. Act like one from the start. Raise your hand to take on additional responsibilities and projects. You'll be surprised at how much you can learn and how many more opportunities will come your way, eventually propelling you into becoming a successful TPM.

PUT IT INTO PRACTICE

✓ Aspiring TPM? Create a transition plan and share it with friends and coworkers who can help you. This will make the commitment real for you. Additionally, someone in your circle might know someone else who can help you or may even have the right role for you. So let the universe know :)

✓ Interviewing? Review and update your resume to pass the six-second rule. Go one step further and do a test with your old and new resume to see if there is a difference in the number of responses.

✓ Understand the competencies tested and start writing down your project examples for each one of them.

✓ Draw out the system architecture for your current programs, and ask questions around tradeoffs, edge cases, etc. to deepen your technical understanding.

Part 3

FOCUSING ON MEANINGFUL IMPACT

Either you run the day or the day runs you.

– Jim Rohn

CHAPTER 13

NAVIGATING YOUR NEW ROLE

Joining a new company is akin to an organ transplant—and you're the new organ. If you're not thoughtful in adapting to the new situation, you could end up being attacked by the organizational immune system and rejected.

– Michael D. Watkins, The First 90 Days

Congratulations! You've landed a new job or been promoted to a higher position. The real journey begins now. Nailing the interview to land this role is just the beginning and the real journey starts now. As you navigate your new role, you will often start at the bottom of the learning curve. You will need to learn and do things you have never done before.

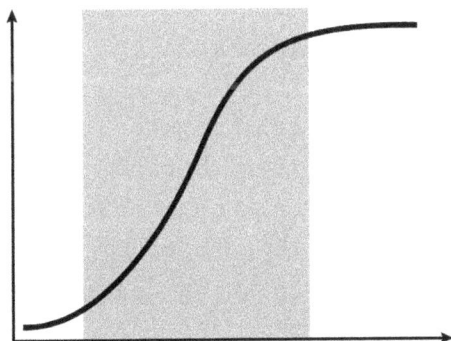

Typical Learning Curve

That's exactly what happened to me when I first joined a big company. The skills I had developed working at small startups prior were not enough. Starting in a new company meant being back at square one. I had to build trust and credibility all over again with a new set of people. In parallel, I had to deliver results, learn about the culture, come up to speed with internal systems and so on. It took me a few months to truly understand the culture, how people operated, and what impact meant in my role. I wish I had understood sooner how to navigate being in a new place. Most companies want to start seeing results from new team members within 90 days, so it is crucial to approach your first few months with intention.

Let's explore five strategies that will help you thrive in your new position and accelerate you towards your next goal.

CREATE A 30-60-90 PLAN

One of the first steps in navigating your new role is to create a 30-60-90 plan. This plan will outline your goals and priorities for the first three months on the job. The 30-60-90 day template is a tool I have personally found to be highly effective when stepping into new roles. When I stepped into the role of a senior TPM leader, I used this template and shared it with an engineering VP. I broke down the plan on how I was going to help the engineering organization get over the challenges and improve execution. Using this approach and sharing it turned out to be remarkably powerful, not only due to its clarity but also because it showcased my structured approach, leaving my stakeholders deeply impressed and confident in my abilities. I then went on to share this template with my team and made it part of the onboarding process. You can use this template or format at any stage of your career even if you are changing teams within the same company.

The first three months in any role is a balancing act between learning about the organization and delivering results. The first three months are broken down in a way to help you start performing or showing results by the end of the third month. The first month is aligning and understanding, and the second month shifts into action mode, so that you are able to deliver on your 90-day goals. Instead of just writing down 3-4 goals for the first 90 days, this format requires you to go deeper into priorities, learnings, specific wins, and your own personal goals, so you can be intentional in your approach.

	0-30 ALIGNING <Focus Area>	30-60 EXECUTING <Focus Area>	60-90 PERFORMING <Focus Area>
PRIORITIES	• Priority 1 • Priority 2 • ...	• Priority 1 • Priority 2 • ...	• Priority 1 • Priority 2 • ...
LEARNING	Goals: • Goal 1 • Goal 2 • ... Output:	Goals: • Goal 1 • Goal 2 • ... Output:	Goals: • Goal 1 • Goal 2 • ... Output:
KEY WINS	Goals: • Goal 1 • Goal 2 • ... Output:	Goals: • Goal 1 • Goal 2 • ... Output:	Goals: • Goal 1 • Goal 2 • ... Output:
PERSONAL	Goals: • Goal 1 • Goal 2 • ... Output:	Goals: • Goal 1 • Goal 2 • ... Output:	Goals: • Goal 1 • Goal 2 • ... Output:

30-60-90 Day Template (Adapted from Deb Liu's Post (Liu 2021))

Here are the basic steps for completing the template.

Identify key priorities: Understand the key priorities that you need to focus on in the next 90 days. Identify which ones can be addressed in the first month versus later so you can sequence accordingly. These priorities become your high-level objectives for the first three months and they are ideally aligned to your job responsibilities and organizational goals.

Example: *A priority might be Program X, which is important for the team. You are coming in to drive that program, and some milestones will need to be delivered within the next three months.*

Focus on learning: Since you are new, there may be tons of context that you require to execute the priorities. Identify what information and context you need along with any names of people who can help you get there. In the case of Program X, you may need to talk to several people to understand the history, how the program is doing, and learn about stakeholders.

Break down your goals for key wins: Divide the priorities or objectives into smaller, actionable goals. This will help you stay focused, measure your progress more effectively, and showcase these as wins at the end of three months.

Example: *With Program X, you may come up with goals like: start driving program meetings by taking over for the manager, or start sending weekly progress reports or address key dependencies to deliver specific milestones.*

Remember to set personal goals: The first 90 days can be a frenzy where you have a ton to do and learn and you are eager to make an impact and prove your worth. However, don't forget about setting personal goals. These can be about building relationships with stakeholders, learning the architecture of the new technical area, etc.

Once you complete the template, seek input on the plan with your manager and key stakeholders to gain their insights and ensure your goals are aligned with the expectations of the organization.

PRO TIP: Stay adaptable and open to change. Your plan may need adjustments as you learn more about your role and the organization.

FIND AT LEAST TWO MENTORS

Having mentors can significantly accelerate your growth and success in your new role. Identifying mentors early can help you navigate challenges that come with a new job and avoid common pitfalls. Consider finding mentors both within and outside your immediate team to get diverse perspectives. Approach potential mentors with humility and a willingness to learn. Invest time in cultivating deeper relationships by seeking their advice and feedback regularly. Remember that mentoring is a two-way street - your output is directly proportional to the amount of effort inputted. Meet at regular cadence and bring a topic or agenda to discuss. Follow up on any action items and close the loop with your mentor. Read more about mentorship and how to best work with mentors in Chapter 22.

PRO TIP: Identify one engineering mentor and one TPM mentor so you get to understand the technical domain as well as the program management culture.

BUILD RELATIONSHIPS

In one of my roles that was meant to be highly cross-functional, I met 100 people in my first 60 days. This really helped speed up my learning and ramp up on the problems, pain points, and key decision makers. Everyone was incredibly friendly and willing to share in depth with me about challenges and what they needed from my team. Investing time in building relationships was one of the best things I did in that job. As I solved some of their pain points, it also helped build my social capital, which proved helpful later when I needed to get their buy-in. Building deep trusting relationships is important at every level and becomes even more crucial as you get to higher levels where your scope will expand, which in turn requires you to lead and influence a great number of people. It is never too early to build these relationships and get to know as many people as you can. Here's how you can start building new relationships at work to enable your progress:

Be approachable: Display a positive attitude, listen actively, and be open to collaboration. Approach others with genuine interest and respect. Ask questions with curiosity and make notes, so you can go back and reference them.

Network effectively: Connect with individuals across different departments and levels within the organization. Attend team meetings, social events, and professional gatherings to expand your network. It can be easy to avoid non-work related activities because you are busy or may not feel as comfortable in a casual social setting. However, the ROI on making connections in an organic and non-forced way is huge.

Be a team player: Contribute to team goals, offer help, and be willing to go the extra mile. Building strong relationships is often a result of being a reliable and supportive team member. This is also a great way to jump right in and learn what is important to your organization.

As you build out these relationships, seek feedback to demonstrate your commitment to improvement and build trust to foster even stronger relationships.

PRO TIP: As you make new connections, request each to provide you with names of at least three other people they recommend you meet. This will exponentially grow your network.

ADAPT TO THE CULTURE

Each organization has its unique culture, and understanding and adapting to it is essential for your integration and success. What works in the context of one organization's culture may not work in another. The same goes for different roles. Your own strengths can be limiting depending on the culture of the new team.

Observe and learn: Pay attention to how things are done within the organization. Observe the unwritten rules, communication styles, and decision-making processes. Gathering data by observation will help you come up with the best solutions that are more likely to resonate with the team members.

Seek cultural mentors: Identify individuals who are well-versed in the organizational culture and can serve as cultural mentors. These individuals can provide insights into the organization's values, norms, and expectations. Establish open lines of communication with them and seek their guidance on how to navigate and succeed within the culture.

Adapt communication style: Adjust your communication style to align with the organization's culture. Pay attention to the preferred methods of communication, whether it's formal or informal, and the level of hierarchy in the organization. Different organizations work differently with email, chat, video, documents, etc., so it is important to mold your own style to fit into the new team seamlessly.

PRO TIP: Don't be in a hurry to make changes and refrain from advising on how it was done in your previous organization. You will not have enough context during the initial days and can rub people the wrong way.

INVEST IN YOURSELF

Even as you settle into your new role, it's crucial to maintain a growth mindset and continuously invest in your personal and professional development. Continue learning and enhancing your skills, so you are always ready to take on whatever comes next.

Identify learning opportunities: Take advantage of any company onboarding or bootcamp offerings during the first 90 days. Stay curious and seek opportunities for professional growth. Identify areas where you want to improve and set goals for your professional development. These goals can be related to acquiring new skills, expanding your knowledge base, or developing leadership abilities. Attend industry conferences, workshops, or enroll in relevant courses to meet your goals and stay up-to-date with industry trends and advancements.

Seek feedback and reflect: Actively seek feedback from your colleagues, mentors, and supervisors. You can let them know ahead of time that you would love to hear their opinion and get feedback. This will set the expectation, give them time to observe you and make the feedback high quality. Use this feedback to make necessary adjustments and grow in your role. Set time aside daily or weekly for self-introspection and reflect on your performance and areas for improvement.

Build a personal support system: Surround yourself with a network of individuals (i.e. your personal board of directors) who support your growth and development. This can include mentors, colleagues, industry peers, or professional communities. Engage in meaningful conversations, share knowledge, and seek advice from this support system. Read more in Chapter 25.

* * *

With these steps, you'll be well-equipped to navigate your new role successfully. Remember, it's an ongoing process, and with each step, you'll gain valuable experience and contribute to your long-term growth and success.

PRO TIP: Add working hours and block off time on your calendar and messaging system to set the right expectations from the start with your team. It is easy to work 50-70 hours per week when you start in a new role or on a new team because you want to make a good impression. However, it is difficult to scale back hours once you have set that expectation with peers. It is easier to add tasks and hours later than it is to remove them.

CHAPTER 14

UNDERSTANDING EXPECTATIONS BY LEVEL

Excellence is being able to perform at a high level over and over again. You can hit a half-court shot once. That's just the luck of the draw. If you consistently do it… that's excellence.

— Jay-Z

During my time at startups, there weren't any so-called leveling guidelines - a document that clearly lays out expectations for a role at a given level like junior, senior, or staff. As an engineer and then as a TPM, I just had to rely on my manager to tell me the next thing they expected me to do, do it, and then hope to get promoted. Tenure alone can no longer get you promoted to the next level. Moreover, competency guidelines give you an understanding of your career path. It gives you the control to manage your own career. A TPM colleague of mine, Maya, identified that the expectations to get to higher TPM levels did not align with her interest in being deeply involved in a specific domain area. For that reason, she made the shift to product management, so that she could focus more deeply on utilizing her expertise.

Most people aspire to rise up the career ladder because they want to challenge and grow themselves. However, without understanding what is required of you to get promoted to the next level, it can be very difficult. As TPMs move up the levels, their responsibilities and expectations increase exponentially, which is also reflected in their compensation. TPMs at higher levels are expected to take on a more significant scope and have a more extensive organizational impact. They are entrusted with managing complex programs and projects, working closely with cross-functional teams, and driving successful outcomes. Decision-making is also not simple - often there isn't a trivial right or wrong answer. TPMs need to provide clear articulation of optionality with pros and cons and other data to help the leadership team make informed decisions and bring in alignment.

Program Scope / Complexity by Level

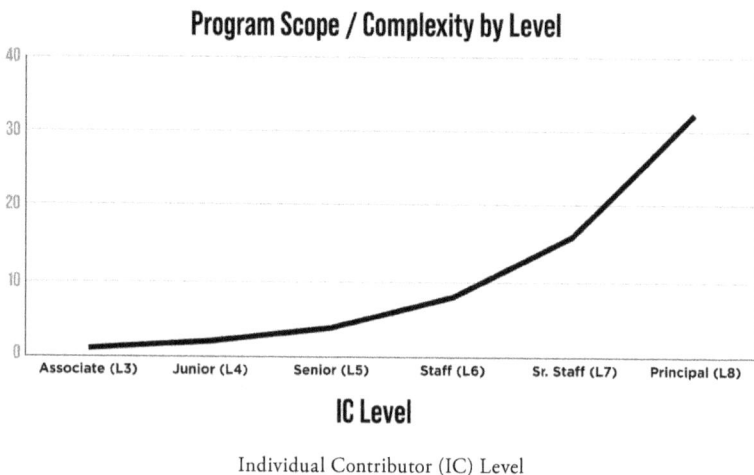

IC Level

Individual Contributor (IC) Level

Most mature TPM organizations will establish a clear set of leveling guidelines to help individuals assess their progress and impact. By understanding the expectations and responsibilities at each level, both TPMs and organizations can align their goals, foster professional growth, and drive successful program outcomes. It also creates a fair performance management system where individuals are evaluated against a rubric rather than ranked against their colleagues.

At a high level, TPM expectations are as follows:

- Break complex initiatives down into prioritized and tangible progress and manage them.

- Ensure consistent flow of information and investment to key stakeholders.

- Identify and manage dependencies and risks proactively across the organization.

- Find long-term solutions to achieve high-quality results.

- Understand team dynamics, product ecosystems, and technical landscapes.

Some organizations - mostly startups - follow a broadband leveling approach with titles ranging from associate to principal. Established companies followed a numbered approach from Level 1 to 10. Most often IC TPM levels start somewhere around Level 3 or Level 4 as the TPM role is generally not favorable for entry-level candidates. That's because this role requires prior understanding of how different functions come together in a product development lifecycle, which comes after working in the industry for a while.

TPMs are generally evaluated on the following competencies, which are similar to the core TPM pillars discussed in the Chapter 5 and interview rubrics in Part 2:

Technical Expertise: The breadth and depth of technical domain and specialized expertise required.

Program Management: The level of scope, complexity, and challenges that can be addressed to create measurable impact. The ability to plan, prioritize, execute, and manage risks to deliver tangible results through creative solutioning.

Communication: The understanding of holistic communication strategy across technical and business teams. The ability to communicate effectively at the right time and to the right audience in a clear, concise, and well-structured manner.

Leadership and Collaboration: The ability to manage stakeholders, influence others, set direction, and make decisions to forward the program and organization.

TPM Levels

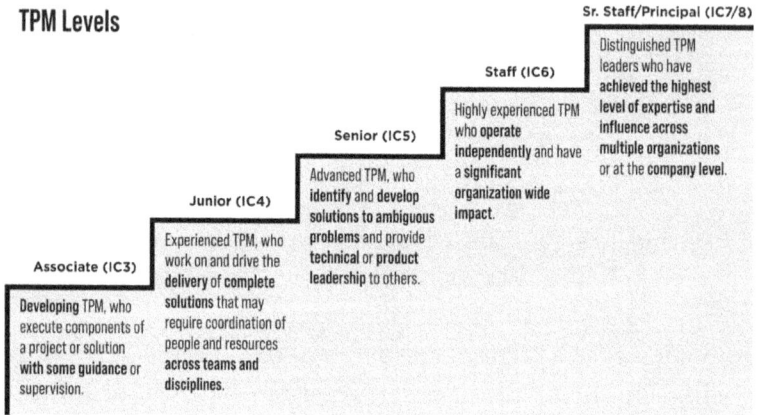

IC TPM Levels Overview

The number of teams that an IC TPM works with increases substantially with each level. These teams may be engineering teams across the technical stack like frontend, backend, infra, cloud etc., or they may be cross-functional like product, design, privacy, etc., or even external vendors.

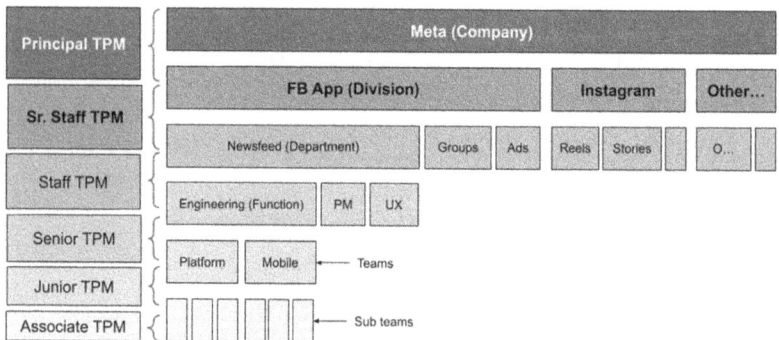

Example of Increasing Cross-Functional Complexity by IC Level

Let's take a deeper look into the expectations for each competency area by level for the TPMs in an IC role. I will use a simple program example of driving a car from Point A to B to demonstrate what a TPM does differently for the same program. Let's call this program "AB."

ASSOCIATE (IC LEVEL 3)

At Level 3, an Associate TPM is a developing TPM who executes components of a project or solution with guidance from a supervisor.

Technical

- Possess a basic understanding of the technical domain and are capable of considering trade-offs between technologies.
- Work on well-defined technical programs where the program or product strategy is defined, but the design and process are not.
- Frequently seek guidance from their manager or senior peers and can represent the work of their assigned team.

Program Management

- Work with ICs to break down cross-functional designs into engineering projects and plan incremental deliverables.
- Plan and execute projects to deliver short-term and long-term commitments in collaboration with multiple teams.
- Identify project blockers and quickly escalate project risk appropriately and quickly mitigate the project's scope or timeline.

Communication

- Produce predetermined documentation, such as execution plans and communication strategies.
- Ensure communication of all project aspects is done via clear, written technical artifacts.

Leadership

- Establish trust with their assigned and partner teams.
- Take ownership of supporting key business outcomes and leverage key partners to align on goals.

On Program AB, an associate TPM gets a map with the entire route plotted out. They are told when to go and who to go with. They successfully drive to Point B by following the directions.

JUNIOR (IC LEVEL 4)

Junior TPMs are experienced professionals who work on and drive the delivery of complete solutions that may require the coordination of people and resources across teams and disciplines.

Technical

- Possess an advanced understanding across underlying systems, services, and related areas and can drive simplification and efficiencies in engineering and complex solutions.
- Represent work of their assigned team, and in-depth project or component status with no overhead to the team.

Program Management

- Collaborate with ICs to break down functional and nonfunctional requirements into cross-functional engineering projects.

- Lead cross-functional collaboration to deliver short-term and long-term commitments, provide updates through reporting, and identify process improvement opportunities.

- Escalate risks or blockers appropriately and propose adjustments to scope or timeline as needed.

- Demonstrate a clear understanding of downstream and upstream impact, identify stakeholders impacted, and proactively manage dependencies.

Communication

- Produce required documentation, manage cross-team coordination and dependencies tracking, and enforce clear and formal communication.

- Identify communication gaps and address them quickly.

Leadership

- Identify needs and collaborate successfully with different customer groups.

- Establish relationships with key stakeholders, and drive key business outcomes.

- Become trusted team members and actively participate in organizational initiatives.

On Program AB, the junior TPM gets the map but only 50% of the route. They figure out who can give them the other 50%, determine who else needs to go, when they need to get to the destination, and if the car has fuel and is in good working condition. With this information, they successfully drive to Point B.

SENIOR (IC LEVEL 5)

Senior TPMs are advanced professionals who develop solutions to ambiguous problems and provide technical leadership to others.

Technical

- Have in-depth technical judgment of underlying systems and services and actively assess how technical implementation choices affect larger goals and trade-offs.

- Assess how technical implementation choices affect larger goals and tradeoffs.

- Actively participate in the design review process, providing understanding of the overall architecture to solve the problem statement.

Program Management

- Lead cross-functional collaboration to deliver desired outputs for stakeholders.

- Contribute actively to process improvements, manage multiple programs with significant complexity, and ensure critical services can hold up to design changes and new features.

- Demonstrate the ability to proactively identify upstream and downstream cross-team dependencies and stakeholders impacted prior to the development work.

Communication

- Communicate complex technical concepts to both technical and non-technical stakeholders proactively.

- Tailor communication style to different audiences and can articulate the value and impact of their work in a compelling manner.

- Produce comprehensive and well-structured technical artifacts and documentation that guide the project teams and serve as references for future initiatives.

Leadership

- Contribute actively to the development of organizational strategy and direction.

- Foster a collaborative and inclusive environment, mentor junior team members, and provide technical guidance and mentorship to others.

- Become recognized as subject matter experts in respective domains.

On Program AB, the senior TPM doesn't get a map because the company dog ate it. After doing a lot of research and talking to several people, they are able to map out the route from scratch. However, the car has some issues, so they need to find someone who can fix it. Once everything is fixed, they drive to Point B and get there on time.

STAFF (IC LEVEL 6)

Staff TPMs are highly experienced TPM who operate independently and have a significant organization-wide impact.

Technical

- Drive large engineering efforts that solve significantly complex or endemic problems. Identifies risks and opportunities in technical strategies, system architecture, and engineering organizational structure.

- Enable other TPMs on the team to be more effective through wide technical judgment.

Program Management

- Lead and drive cross-functional initiatives with a focus on strategic goals and objectives.

- Have a strong grasp of program and project management methodologies and be able to adapt them to suit the needs of the organization.

- Manage resources, budgets, and timelines effectively while aligning efforts across multiple teams and stakeholders.

- Propose new development frameworks to increase effectiveness and productivity at the department level.

- Ensure risks at department level are identified early during planning and throughout development in a timely manner.

- Identify and mitigate risks at a strategic level, considering long-term implications and potential business impacts.

Communication

- Influence and align stakeholders at all levels of the organization.

- Present complex technical information to executive leadership, steering committees, and other key decision-makers.

- Leverage deep domain knowledge to articulate the strategic value and impact of their work, fostering support and buy-in from stakeholders.

Leadership

- Play a key role in shaping the organizational culture and driving innovation.

- Mentor and develop junior TPMs, provide guidance and support to teams, and actively contribute to the professional development of the broader organization.

- Become trusted advisors and get involved in strategic planning and decision-making processes.

On Program AB, the Staff TPM is not convinced about the reason to go to Point B. After discussing this with program stakeholders, they all agree that Point C has more benefits than B and is also 25% closer. They then map out the route and decide to use a different car because it's more fuel efficient. They get to Point C well before the deadline.

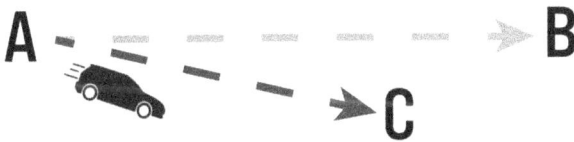

SR. STAFF/PRINCIPAL/FELLOW (IC LEVEL 7/8)

Principal TPMs are distinguished leaders who have achieved the highest level of expertise and influence within the organization. They are recognized as industry experts and thought leaders in their respective domains.

Technical

- Provide strategic vision and direction, shaping the organization's technical roadmap and driving innovation at scale.

- Understand and coach the team on engineering excellence. Develop teams and other TPMs in their domain.

- Have a wide system view across the company and ability to work with engineering leaders to continuously simplify architecture.

Program Management

- Drive large-scale, transformative initiatives and programs.
- Have a holistic understanding of the organization's strategic goals and align programs to achieve those objectives.
- Work closely with executive leadership and senior stakeholders to ensure the successful delivery of key initiatives.
- Identify and address complex, high-impact risks that have the potential to significantly impact the organization.
- Identify ways to disambiguate unknown risks at the department/company level.

Communication

- Influence and inspire at the highest levels and effectively communicate complex technical concepts to diverse audiences, including industry conferences, executive briefings, and board meetings.
- Become skilled storytellers who can articulate the vision and impact of their work, fostering support and driving organizational alignment.
- Identify collaboration gaps and escalate where needed to improve development frameworks.

Leadership

- Provide mentorship and guidance to senior leaders and executives.

- Contribute to the development of the organization's technical talent, foster a culture of innovation, and lead by example.

- Serve as trusted advisors to the executive team and provide strategic guidance on technical investments, partnerships, and industry trends.

- Participate in shaping the organization's long-term strategy and be instrumental in driving innovation and technological advancement.

On Project AB, the Principal TPM is perplexed and wondering if a car is the best way to get to Point B. It gets stuck in traffic, isn't scalable to carry more people, doing multiple trips to different places, etc. They propose alternatives to the program stakeholder and teams like an elaborate train system that reaches Point B plus many other places of interest, making it a versatile and scalable option in the long term. They have to convince the business teams, and get buy-in from multiple leaders as they need more resources. They work with multiple teams to finalize solution scope, navigate the development issues, and eventually launch the system. Everyone is able to travel more efficiently in the future.

The expectations laid out for each level are not comprehensive by any means. Specific responsibilities of TPMs may vary across different organizations and industries. This provides a general framework for understanding the requirements for career progression. The levels may not necessarily always correspond to the job titles. Many smaller companies tend to provide higher-level titles to attract talent. Investment banks have Assistant Vice President (AVP) and VP titles for mid-career folks and Directors are actually above VP. Even Salesforce regularly uses Director titles for IC Level 6 scope of work. Higher titles are generally used to provide authority in interaction with external vendors or partners.

It's always a good idea to consult the specific career paths and frameworks within your organization or industry to get a more accurate understanding of the TPM roles and responsibilities.

PROMOTION READINESS

You can use these leveling guidelines to identify the areas where you are doing well and where you need to grow. This will help you identify what programs will give you the opportunity to develop those skills.

Competency Pillars	Area	Current Level	Next Level	Examples	Opportunities
Technical Knowledge & Judgment					
Program Management	Scope				
	Execution				
	Risk Management				
Communication					

Promotion Readiness Template

Here's what I did with my team and something you can do as well if your team has a list of requirements for each level laid out.

- Take the requirements of the current level and the next level and put them in two columns side by side. Add two more columns to the right - see the promotion readiness template.

- Now mark each requirement or expectation under both columns. Mark the cell with these three colors: green - doing well consistently; yellow - doing well sometimes or inconsistently; and red - not doing well at all or not good at it.

- Ideally for a strong promotion case, you should have all greens at the current level and many greens, some yellows, and minimal red at the next level. This indicates that you may be ready for a promotion.

* * *

Note that any leveling guidelines are not meant to be used as a checklist. Promotions within the organization are based on multiple factors such as achieved results, demonstrated behaviors, and the company's current business. In many companies, promotions are lagging and you need to be performing at the next level to be promoted. This is to ensure that you succeed at the next level as well because you already have experience handling the scope and intensity at that level.

PRO TIP: Keep a running list of your accomplishments along with details on how you achieved those results. This will help you quickly reference your past achievements when building a case for promotion.

CHAPTER 15

BEING INTENTIONAL ABOUT GOALS

*People with goals succeed because
they know where they are going.*

– Earl Nightingale

How often do you use Google or Apple Maps when you get into your car? I remember the first time I started using GPS - one of those Garmin devices. It proved immensely helpful in planning out my route, especially when I moved to a new city. Most importantly, I didn't have to look at minute details on an atlas and I didn't get lost. Now, even though I am familiar with my journey, I still use Maps because it gives me information like traffic, road closures, etc. that are helpful for me to make real-time adjustments.

Just like mapping out your route before getting into a car, setting goals helps you be more intentional and strategic in your journey. You often work with your team to develop project or program goals. However, it's equally important to apply the same principles to setting personal and professional goals. In this chapter, we will dive into the process of setting SMART goals for TPMs, ensuring they are Specific, Measurable, Achievable, Realistic, and Time-bound. By doing so, you can navigate your path with clarity, focus, and effectiveness. Setting goals helps you be more mindful and protects you from getting bogged down by day-to-day fires or tactical requests. You should set goals for both professional and personal aspects of your life.

The Power of SMART Goals

How many times have you set New Year's resolutions and not kept them? According to multiple studies, 80% of New Year's Resolutions get abandoned by mid-February. *In one study, 35% of participants who failed their New Year's Resolutions said they had set unrealistic goals. In another, 33% of participants who failed didn't keep track of their progress. Yet another study mentioned that 23% forgot about their resolutions, and about one in 10 people who failed said they made too many resolutions* ("New Year's Resolution Statistics (2023 Updated)", n.d.).

New Year Resolutions are not SMART! No wonder so many people give up on them in a couple of months! They fail because they lack specificity, achievability, and time-bound targets. It's crucial to understand the significance of setting SMART goals rather than relying on vague resolutions or general aspirations. By setting SMART goals, you can overcome these limitations and ensure that your goals are actionable, practical, and aligned with your desired outcomes. With numerous responsibilities and competing demands, prioritization of goals can also help to focus your efforts on what will help achieve the greatest impact and contribute to your overall success.

CRAFT SMART GOALS

The beginning of the year is the most common time to craft your goals and align them to the top level company and team goals. Sometimes setting goals for the entire year seems daunting because you may not know how things will change. In that case, I recommend you set goals at least every six months for yourself and do a mid-cycle check-in. That means set goals in January and again in June or July. Track your goals every month, but do an intentional check in April/October. Here are six steps to craft your SMART goals:

- **Define impact:** Start by thinking about what is most meaningful for you to achieve in a defined time-period and the underlying impact. Identify the "why" behind your goals and how they can contribute to personal or organizational growth. This can align your goals with a sense of purpose and motivation. Jot down bullet points to create the first draft.

- **Be specific:** Specificity is key when setting goals. Set clear, well-defined goals that are actionable and aligned with your desired impact. In the professional setting, you can think of impact as a joint outcome and reward that the entire team gets and the goal is individual action completed to meet the impact. You can have more than one goal per impact.

- **Prioritize your goals:** Once your rough draft is done, prioritize your desired goals into must haves and nice to haves or prioritize by ROI versus Effort. Limit yourself to 3-5 main goals that are aligned with the organization. This way you can stay focused and do a great job at each rather than a mediocre job trying to do too much.

- **Align with competency guidelines:** Reference the expectations at your level to ensure that your goals are providing you the opportunity to demonstrate your expertise appropriately for each competency at your level.

- **Define success criteria:** To track your progress and evaluate the success of your goals, it's crucial to establish measurable success criteria. Define concrete metrics or indicators that will allow you to assess your achievements and stay on track. Ensure that your success criteria are not just about finishing a set of tasks as those do not necessarily indicate tangible results. Metrics also help you quickly identify if you have met or exceeded your targets or expectations. For each goal that you prioritize, ensure that you are focusing on the impact and have a clear method to measure success.

- **Align your goals:** Connect with your manager to align on your priorities and ask for feedback. Utilize any specific format or framework that the organization is using, so that you are speaking the organization's language.

Example: Some companies use OKRs and every team member aligns their own OKRs with the next level up.

Here's an example of a SMART TPM goal:

Goal: Define a new process to provide more visibility and predictability to engineering leadership. Implement and get adoption of this new process with at least two teams in Q1.

See how this goal doesn't end after the first sentence? In fact, it goes further to define the success criteria.

Setting goals that are both achievable and realistic is essential for maintaining motivation and momentum. Assess your capabilities, consider potential obstacles, and adjust your goals accordingly to ensure they are within reach.

DEVELOP ACTION PLAN

Setting goals is only the first step; developing action plans or the "how" is equally crucial. Identify specific activities and tasks that will contribute to achieving your goals. By establishing a clear connection between actions and outcomes, you can chart a practical path towards success. While the list of tasks can be long, ensure that the ROI on each task is high.

Once you have outlined 1-3 impact areas and 3-5 goals, reference the leveling guidelines in Chapter 14 to anchor yourself to ensure that you are not skewed towards any one competency area. Have a good mix of programs that leverage all your TPM skills for a well-rounded performance.

Example: *If you only lead programs that utilize your execution chops, but don't give you an opportunity to showcase leadership skills, it may not help you get to the next level.*

IMPACT 1:

[The why/vision or joint outcome]

Goal

[Your specific goal to achieve the impact from a TPM standpoint. The number of goals per Impact should be between one and three. This ensures you are not trying to do too much]

Success Criteria

[How will you measure success]

How will I achieve this goal?

[What will you do across the 3 TPM axes (Technical, Program Management, Leadership)]

Impact 2 and so on...

[Repeat the format for each Impact area. Limit to three]

Goal Setting Template

REVIEW AND FINALIZE

Before finalizing your goals, take the time to read your document multiple times. This rumination process allows you to approach it with fresh eyes each day, enabling you to spot areas that may need refinement. Trim unnecessary details, clarify ambiguous language, and make sure your goals are crisp and concise. Consider the overall coherence and alignment of your goals, ensuring they collectively contribute to your desired impact.

To gain additional perspectives and insights, reach out to someone you trust—a friend, mentor, or colleague—and ask for their feedback. They can provide valuable suggestions and help you strengthen your goals. Incorporate their insights into your document, refining it further based on their feedback.

SHARE YOUR GOALS

Goals shouldn't be kept solely between you and your manager. It's crucial that those you interact with, including team members, stakeholders, and collaborators, understand your goals as well. Sharing your goals with others creates a sense of accountability and allows them to offer support, guidance, and collaboration. You never know who might have valuable resources, insights, or connections that can assist you in achieving your goals.

By sharing your goals, you become more transparent and open to opportunities for growth. Others may be inspired by your objectives or provide valuable input that helps refine your approach. Remember, accountability is not limited to one person—it can come from various sources, and the more people you share your goals with, the more invested they become in your success. Additionally, sharing your goals with a broader audience contributes to the power of manifestation. By verbalizing your intentions and putting them out into the universe, you invite the energy and resources needed to support your endeavors. The act of sharing your goals amplifies your commitment and belief in their achievability.

* * *

Your ability to set and achieve SMART goals contributes not only to your personal growth but also to the success of your programs, teams, and organization. Embrace the power of SMART goals, and let them guide you on your journey towards continuous improvement and exceptional performance as a TPM.

PRO TIP: Assign specific timelines or deadlines to your goals, allowing you to stay accountable and maintain a productive pace. Time-bound goals create a sense of urgency and focus.

CHAPTER 16

SCALING YOURSELF TO LEVEL UP

Don't confuse being busy with being productive.
Just because you're busy all the time
doesn't mean you're getting closer to your goals.

– Karen McKenna

In my early days as a TPM, I often wanted to know all the technical details because I thought that's how I could better manage a program. If I was assigned to a program halfway finished, it would take me time to ramp up on what had already happened. In order to learn, I would get into the weeds too much, which meant I was sometimes missing broader signals and could not always ensure that things didn't fall through the cracks. As I got more experience, I learned to get over the discomfort of not knowing everything from the start. It also meant that I had to figure out another way to get the information that was just right enough to assess if I needed to dive deep. This allowed me to scale myself and balance the breadth and the depth as well as the technical versus non-technical needs of a program. Today, I hear many TPMs having the same feelings and I emphasize that at some point, knowing every single detail is not conducive to your long-term goals of leveling up and managing big programs.

You are driving multiple programs, juggling competing priorities, navigating complex dependencies, and communicating with different teams. You are striving to balance the breadth and depth of your work. You are also expected to manage bigger scope at senior or higher levels. Today, you may be working with three teams and can attend everyone's standup meeting. However, tomorrow, if your program runs across 10 teams, you do not have enough time in the day to attend everyone's scrum meetings. For every 20-40 engineers, there is maybe one of you. This means there will always be more work than the time you have.

You need to find another way because you can't work endlessly round the clock. You have to choose how you work, what balls to drop, and when or to whom you can say no. Your ability to scale becomes essential for achieving professional growth. It is also what differentiates a good TPM from a great TPM. Scaling yourself means you can drive highly complex initiatives with ease without working overtime. In order to do so, it's essential to break free from the scrum master mode that may hold you back from reaching full potential. By being intentional and leveraging other people or systems, you can multiply your impact while keeping the effort required linear.

FOCUS ON STRATEGIC EXECUTION FIRST

As I grew in my role and started managing a bigger scope, I was working with multiple teams. I could not attend every team's standup or update their tasks. So I worked with my engineering peers to split responsibilities to ensure that the team was still executing and I was supporting the cross-functional aspects while still keeping on top of every team's progress. Over the years, I helped TPMs on my team do the same. I aligned with engineering to help clarify how execution was managed at different TPM levels and where we need engineering support. The call out to strategic versus tactical execution came out of one such discussion where I wanted to provide a clarity on where TPMs can add value, especially at higher levels.

I like the term "strategic execution" because it emphasizes the key role that a TPM plays in a program. It shows that the TPM role is a thinking role, where the TPM has to come up with a robust program execution strategy that will create success for the long term. A TPM cannot be just handed over a program where they check off a list of tasks. The heart of the TPM role lies in strategic execution, which is crucial to becoming a successful TPM.

The terms "strategic" and "tactical" are not commonly associated with execution in program management. Strategic thinking involves identifying long-term goals and planning the means to achieve them. Strategic execution similarly means that a TPM is identifying and monitoring the holistic view of the program - from risks to dependencies to what every stakeholder needs and wants. This goes beyond the tactics of running the program itself like creating sprint boards, writing status reports, or running a meeting. Here are some ways to distinguish between strategic versus tactical execution:

STRATEGIC	TACTICAL
Developing a robust execution plan for your program	Identifying tools to track work
Creating a roadmap with timeline and milestones	Creating project structures and tasks in tools like Jira
Breaking down your timelines into smaller deliverables	Conducting sprint planning, backlog management, retrospectives, etc.
Identifying the dependencies and risks and creating mitigation plans	Facilitating daily standups, getting updates from team, and checking off tasks
Identifying key stakeholders and building relationships	Helping core team in their work like creating/updating tasks
Creating a communication plan and knowing your audience	Reporting on weekly basis

Strategic vs. Tactical Execution

In any program, there is a mix of strategic and tactical activities. It is essential to strike a balance between these two types of executions to ensure alignment with long-term goals along with prioritization of the immediate term tasks. Both strategic and tactical activities are required and take up significant time. To ensure meaningful impact, try to spend the majority of your bandwidth on high-impact, strategic execution activities.

PRO TIP: Set aside 15 minutes at the end of each week day and 30 minutes end of Friday to reflect back on the day/week. Assess how your actions throughout the day lead to your goals. Plan how you want to be strategic the next day/week. This will ensure that your days don't pass by in a flurry of tactical activities such that you lose sight of the long-term goal.

PRIORITIZE RUTHLESSLY

In one of my startup roles, we used to create a stack ranked program list called "Forced Rank." We were trying to say that just because a program appears at number 10 on a list doesn't mean it is less important. It just means that there are nine more important things to do. If the team has the capacity to do all 10, then that's great but if not, then it's okay to take it up later.

Ruthless prioritization is a term often used by Sheryl Sandberg, the former Chief Operating Officer (COO) of Meta/Facebook. She always said, *"Ruthless prioritization means only focusing on the very best ideas. It can get hard because you're always trying to do more, but it's one of the best and most important ways to stay focused"* (Sahadi 2018). There will always be competing priorities and they may all be positive for your business, but that doesn't mean you have to do all of them.

SCALING YOURSELF TO LEVEL UP | 163

Just like product features or project lists, your own work also needs to be prioritized and you need to force-rank your tasks based on factors like cost-benefit analysis, ROI, and scalability.

If it doesn't hurt, you haven't prioritized hard enough!

You possess a unique talent for gracefully juggling competing priorities and seamlessly switching contexts. Your ability to handle multiple programs simultaneously is commendable.

However, multiple studies have confirmed that true multitasking—doing more than one task at the same time—is a myth. People who think they can split their attention between multiple tasks at once aren't actually getting more done. In fact, they're doing less, getting more stressed out, and performing worse than those who single-task (MacKay 2019). According to a Harvard Business Review article, there is a cost to context switching - it's two seconds per switch which adds up to five weeks per year (Murty, Dadlani, and Das 2022). So, in reality, you can't effectively do your job if you have too many priorities on your plate. There inevitably comes a point of diminishing returns when attempting to manage an overwhelming array of tasks. Recognizing and understanding this tipping point is essential, as it varies for each individual. Driving 1-3 medium- or large-sized programs is better than coordinating 10 small projects. The smaller projects don't give you enough opportunities to develop key skills as you end up spending a lot of time context switching or performing tactical and superficial tasks. If you empower technical leads to run the semantics on smaller projects, you can still drive and maintain the overarching portfolio as a TPM lead, which can be highly impactful.

To do two things at once is to do neither.

– Publilius Syrus

Learn to Say "No"

In many organizations, there is a prevailing perception that saying "yes" to everything or working long hours is a mark of dedication and success. However, this belief can lead you down a treacherous path, tempting you to take on more than you can feasibly handle. The busyness may create a false sense of importance and value, but the ultimate result is often a series of negative consequences. Missed deadlines, low-quality outputs, and poor communication become an unfortunate reality when you overcommit and spread yourself too thin.

To break free from this cycle, it is imperative to learn the art of saying "no" strategically. By consciously choosing to focus on a few key priorities, you can allocate the appropriate time and resources needed to ensure high-quality outcomes. This intentional narrowing of your workload creates valuable space for deep thinking and concentrated effort, leading to better results for your programs, team, and organization as a whole.

PRO TIP: Drop the right balls - the ones that just keep you busy but don't result in impact or career growth. Take a look at the career level requirements in Chapter 14 to assess what work can help you meet the expectations (especially any red or yellow items). Try to drop the rest.

It is important to have a good judgment of when and how to say "no." Recognize that you will always have new requests coming in and you have to evaluate each of them separately to make trade-offs. Thoughtfully analyze the whole situation - what's incoming and what's already on your plate - and then consider both the immediate and long-term impact. By carefully weighing the value and alignment of incoming requests with your strategic objectives, you can make informed decisions about which opportunities to pursue and which to decline.

Example: If there is a site event which results in a customer issue, you know that getting it fixed is critical. In that case, you will say yes, inform your manager and other key people about it, and you may need to stop working on other priorities for a few hours. On the other hand, if you are requested to provide a report with data for an upcoming customer meeting, you need to assess the priority of this request. You can ask when the meeting is, what the overall objective is, and when they need the report. Based on this information, you can either point them to an existing resource, point them to another person who is better aligned, or take it on but schedule it because you know it's not urgent.

Urgent vs. Important

Incoming requests can feel urgent, especially if they are coming from someone in your management chain or an influential person. However, a constant sense of urgency just makes you reactive, which takes away from creating meaningful impact and may lead to negative consequences. Focus on the important activities to achieve your goals while balancing the immediate demands wisely.

- **Urgent** activities demand immediate attention and are usually associated with achieving someone else's goals. They are often the ones we concentrate on and they demand attention because the consequences of not dealing with them are immediate.

  Example: Meeting a program deadline or fixing a client-facing issue.

- **Important** activities have an outcome that leads to us achieving our goals, whether these are professional or personal.

  Example: Creating a plan for an upcoming program or setting up a dashboard for execution.

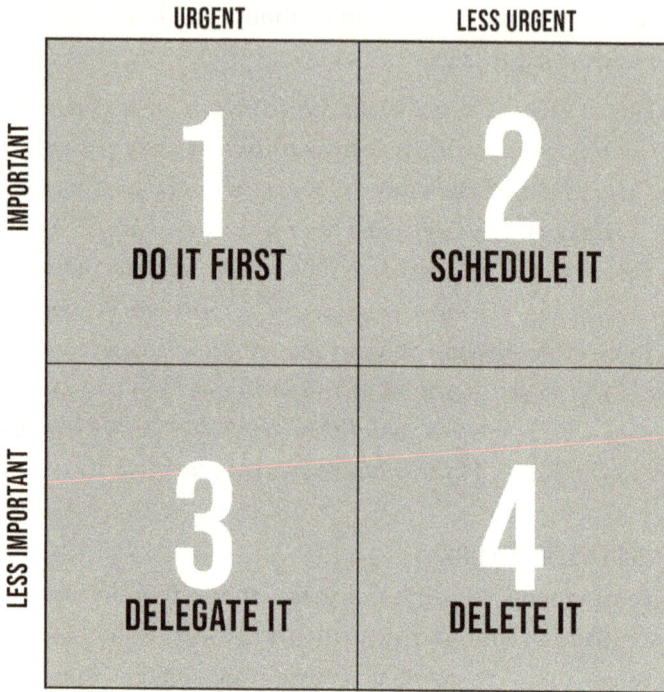

Eisenhower Matrix: Urgent vs. Important

Use the Eisenhower Matrix (Asana 2022) to guide your prioritization decisions.

DO: Urgent and Important (Quadrant 1): Tasks that require immediate attention and are crucial to the success of the program. These should be addressed promptly to prevent problems or mitigate risks.

Example: Resolving critical system outages, addressing security breaches, or handling high-priority stakeholder escalations.

SCHEDULE: Important, But Less Urgent (Quadrant 2): Tasks that contribute to long-term goals and success but are not time-sensitive. Planning, strategizing, and skill development fall into this quadrant.

Example: Developing a long-term technology roadmap, implementing process improvements for efficiency, or investing in team skill development.

DELEGATE: Urgent, But Less Important (Quadrant 3): Tasks that demand immediate attention but don't significantly contribute to long-term goals. They can often be delegated or minimized to focus on more crucial matters.

Example: Attending routine team meetings that could be delegated or handling minor system glitches that another team member or engineer can tackle.

DELETE: Not Urgent and Less Important (Quadrant 4): Tasks that are neither urgent nor as important. These should be eliminated or minimized as they don't add value to the program.

Example: Reviewing trivial reports or engaging in excessive administrative tasks.

Effectively utilizing the Eisenhower Matrix allows a technical program manager to focus on high-priority activities critical for success while planning for the future and managing day-to-day urgencies efficiently. It provides you a clear criterion for decision-making to effectively communicate your focus areas to your stakeholders and keep them informed of any changes.

Work Yourself Out of a Job

If there are tasks and activities on your plate that are not conducive to your long-term goals, then you don't have to keep doing it. When I started out as a TPM at a big tech company, I got the opportunity to be at the center of an important initiative that involved 30+ teams from across the organization. I was in multiple meetings every day and getting involved in various activities. A few weeks later, I felt uneasy because I knew I was doing important work but something was missing. I wasn't enjoying much of the work. At the same time, I attended a training that talked about a strengths-based approach based on the book "Standout 2.0" by Marcus Buckingham where he asks you to make a list of work things you love and loathe. I made a list of everything I did in a week and it was so helpful to see where I was spending my time. In fact, it was a shock because I didn't realize that a disproportionate amount of my time was going to mundane tasks that were not reaping any benefits for my growth. I took that list and prioritized the aspects that would help me scale.

> **PRO TIP:** Aim to put yourself out of your current job - the aspects of your job that can be automated and/or delegated. Think of your time as money. What is the ROI on your investment of time? What else could you be doing that will help you reach your goals?

Here's how you can start scaling yourself:

- Start by examining your weekly tasks, everything you do from meetings to reports to tasks.

- Categorize them based on their value, manual nature, and level of monotony.

- Use the Urgent versus Important context or another prioritization framework to identify opportunities for automation, deprecation, or delegation. For instance, you may no longer need to maintain a dashboard because a specific phase is over and the information is no longer relevant.

- Automate where possible - as TPMs you are great at optimizing, so use those strengths along with technical skills to automate tasks that you can't delegate for whatever reason. That way, you are reducing work for even other folks on the team.

Remember the saying, "If you give a (hu)man a fish, you feed them for a day. If you teach them to fish, you feed them for a lifetime." Don't worry about how to delegate - start by building trust and emphasizing mutual benefits. Share your knowledge and expertise with your team, so they feel empowered to take ownership. By teaching others, you can eliminate yourself as the bottleneck and help achieve results faster. By doing so, you are also gaining valuable mentorship experience. Where possible, set up lightweight systems or processes for engineering teams. Create repeatable processes for yourself that remove low-value tasks, allowing you to concentrate on strategic work. Each repeatable process will allow you to free up time to focus on high-value work like managing cross-team dependencies. It will help you build credibility as others will leverage these simplified workflows for their own benefit.

Examples:

- *You need to generate a progress report weekly. What parts of the report can be automated through dashboards? What tools can you utilize to collect and display data or run queries every week?*

- *You need to monitor certain topline metrics and send emails to key stakeholders if they drop below a threshold. Can you set up alerts to notify you instead of you having to go to the metrics dashboard? Can you automatically send an email to someone who can take action?*

- *Creating an on-call process that can be rotated through the teams, so no one person will bear the responsibility for eternity. It will help create redundancy and ensure you have strong backup mechanisms.*

- *Train engineers and technical leads how to navigate sprint boards or run a scrum, so you can find time to track dependencies and mitigate critical risks with key stakeholders.*

The ability to focus on fewer priorities requires both self-awareness and a strong understanding of your team's capabilities and capacity. By honestly assessing your own limitations and setting realistic expectations, you can avoid overcommitting and ensure that your attention is dedicated to the programs and initiatives that truly matter. This self-awareness also empowers you to effectively communicate with stakeholders, enabling them to understand the reasoning behind your choices and the implications for overall success.

BUILD ACCOUNTABILITY IN OTHERS

TPMs can often use deadlines as a way to hold others accountable. However, sometimes you may not have such privilege at your disposal. Holding others accountable means that they are clear of their expectations and they will hold themselves responsible for the outcome of their deliverables. Once you create those prioritized lists and start delegating tasks, you will need other folks to deliver in a timely manner. You may not always have the time to chase them down, but your success still depends on those tasks getting done when you need them. If you can instill an accountability culture in your team, your job as a TPM will become easier.

Here are five steps to help you build accountability with the team:

1. **Common purpose:** Why does this matter and why them?

2. **Clear expectations:** What does success look like for each person and what are the timelines and definition of done? Establish a culture of trust and avoid micromanagement.

3. **Communication and alignment:** How do we set the overall team or task for success? Encourage proactive reporting of progress and problems.

4. **Collaboration and guidance:** How's it going and are any adjustments/help needed? Regularly check in with your team members and assure them that you are there to help if they run into any issues, so that they can come to you quickly.

5. **Celebrate and calibrate:** Celebrate their accomplishments and thank them for being responsible. What have you learned from any setbacks and how will that drive future actions?

PRO TIP: In programs involving dependencies on multiple people on multiple teams, liaison with one specific individual from each team to manage deliverables for their group. This will help you not have to check in on every person on that team. Creating a culture of accountability is also important for people managers and leaders as their results are dependent on others. Accountability provides ownership, which leads to high-output teams and strengthens your leadership abilities.

ENSURE MEETING EFFECTIVENESS

Post pandemic, with remote and hybrid work becoming commonplace, most workers are tired of being in meetings. Meeting overload is a real problem that has not been fully solved yet. Meetings should be set up only for necessary discussions that require people to be together at the same time. If something can be resolved asynchronously, then a meeting should never be required. Prioritize asynchronous decisions and outcomes by providing a clear framework.

At the same time, there will always be some meetings required in order to move your program forward like roadmap planning, program risk assessment, etc. As a TPM, you will lead these meetings, but you have to be really mindful that every meeting is effective and worth the time of all its attendees. It is important to set clear meeting objectives, manage meeting dynamics, and ensure productive collaboration and decision-making. This will help you drive alignment, share information, and make decisions collaboratively with less meetings. Well-facilitated meetings save time, enhance team productivity, and foster engagement among participants. Here are a few tips to make meetings more effective and, in turn, reduce the number of hours spent in meetings:

Plan in advance: Take time to define clearly the desired outcomes for each meeting. Is this meeting to make a decision, review strategy, brainstorm ideas, or something else? Share the agenda with participants and set expectations around any advanced preparation and attendance.

Trim the attendee list: It is easy to invite everyone to a meeting, but take time to include only those needed to help you get the desired outcome. If you need a decision maker, then include them. If you need someone's opinions, invite them. However, if someone just needs to be informed, they don't have to be in the meeting. Follow up with them offline. Smaller meetings are often more effective.

Manage meeting dynamics: Set expectations for participation, establish ground rules, and ensure that all voices are heard. Ensure that discussions are leading to the desired outcome and redirect conversations when necessary to maintain focus and productivity.

Encourage active participation: Foster an environment where all participants feel comfortable contributing their ideas and perspectives. Use inclusive facilitation techniques such as round-robin discussions, brainstorming sessions, and breakout groups to encourage active participation.

Track and follow up on action items: Assign action items, document decisions, and track progress during the meeting. Follow up after the meeting to ensure that action items are implemented, and communicate updates and outcomes to relevant stakeholders. This way, everyone understands the value of the meeting and anyone who wasn't there still gets the desired information.

Continuously improve meeting effectiveness: Regularly solicit feedback from participants to identify areas for improvement. Adjust meeting formats, cadence, and facilitation techniques based on feedback to optimize engagement and productivity.

Cancel meetings if needed: If you determine that a meeting is no longer needed or that the outcome can be achieved offline, then don't be afraid to cancel the meeting.

PRO TIP: Encourage others to follow the meeting effectiveness etiquette. This will help create a broader culture and reduce FOMO (fear of missing out).

UTILIZE THE POWER OF AI

Taking meeting notes is one of those tasks that most people dislike and TPMs often become de facto note takers. However, it can be difficult to drive a conversation effectively while taking down notes. Well, now we don't have to go around the room taking turns jotting notes.

AI is no longer the stuff of science fiction. It's a reality that's transforming industries, including program management. AI can be a total game changer for TPMs because it will enable TPMs to scale effectively. AI will help TPMs focus on strategic execution while it handles the tactical aspects automatically. This means you get more time to focus on mitigating risks, building relationships with stakeholders and communicating more effectively across different audiences.

You can take advantage of AI for more than just note-taking. AI can be helpful with many day-to-day TPM activities:

Communication: AI can help summarize action items from those notes and send them out to relevant stakeholders. You can use AI to automate follow-up reminders on those actions items, update, or mark them as complete. You can use existing integrations in your email, messaging, and video conferencing tools or build them for yourself.

Reporting: Providing regular program status updates to stakeholders is a core part of a TPM's work. However, creating reports that cater to different levels of audiences and align with company culture can be a daunting task. You may also need to create multiple reports for different audiences or formats. You can reduce the time spent creating these meetings by using AI to help you structure your reports in a way that meets the objective. This means you can focus on adding in data and your insights while the language structure, etc. is taken care of. While you can use any tool to do so, you will need to experiment a bit at the beginning to make sure that the report being generated sounds like you and meets the goal of your stakeholders.

Execution: Estimation, resource allocation, risk management, and quality assessment are all critical steps of program execution that rely on multiple levels of data - from requirements to historical context. AI can speed up many of these activities by doing a predictive analysis based on past and present information. It can also help speed up the process and reduce human error or bias.

Explore what AI can do for you to help you scale. While there are many existing tools and apps, more advanced tools and apps are being built, so do you research to understand what would work best for you. Stay updated and be willing to learn and adapt are the keys to success in this AI-driven era of program management. AI will help TPMs scale and focus on higher value work, leading to better job satisfaction.

* * *

PRO TIP: Have patience as you learn to scale yourself. It can be uncomfortable in the beginning and requires time and practice. Continue to build the muscle for strategic execution by taking on more significant and challenging programs.

CHAPTER 17

ELEVATING YOUR CRAFT

*Hustling brings a dollar for today. Mastering your craft
brings wealth for a lifetime!*

– Unknown

Picture yourself dedicating days upon days to meticulously
designing a process aimed at enhancing your team's productivity.
Every conceivable pitfall is considered, and the perfect solution is
devised. With excitement and enthusiasm, you present the well-
crafted process to your team, anticipating their embrace. However,
as implementation begins, it quickly loses traction, and your team
fails to fully embrace the fresh approach. Eventually, you come to
the realization that your tireless efforts had little positive impact
beyond the valuable lessons learned.

Over the past decade, there has been a shift to focus on
impact rather than just effort. While the concept of impact may
not immediately resonate when you first encounter it, it's essential
to recognize that high impact results from consistently delivering
outstanding work in a remarkable way. Once you have established
yourself, it becomes crucial to elevate your skills in creating
meaningful impact, both for your personal growth and for the
betterment of the organization you're dedicated to supporting.

The Power of Impact

The business impact you accomplish outweighs the sheer effort you put in. In fact, the level of effort is not necessarily proportional to the level of impact. When you understand the power of impact, it becomes painfully clear that spending countless hours on an activity is futile if it fails to yield the desired results. To achieve high impact, focus your most precious resources—time and energy—on the right things. Each moment of your valuable time should be treated with the utmost respect, for it is a finite and non-renewable asset. Be intentional about where you direct your efforts and ensure they align harmoniously with your strategic objectives.

The power of impact lies in your ability to analyze the activities that drive real change and prioritize them. Rather than spreading yourself thin across various tasks, concentrate your efforts on initiatives that will yield substantial results. This approach requires a deliberate and strategic mindset, an unwavering commitment to identifying the areas where your influence will make the most significant difference.

$$\text{OUTCOME} \geq \text{EFFORT}$$

Recognizing the value of collaboration and aligning your efforts with the broader goals of your organization will further amplify your impact. By understanding the larger context and working synergistically with other teams and stakeholders, you create a ripple effect that extends beyond your immediate sphere of influence.

EMBRACE END-TO-END OWNERSHIP

Your role goes beyond mere execution and tracking of engineering work. It extends into strategic thinking and requires you to embrace end-to-end ownership of programs. By doing so, you gain a comprehensive understanding of the big picture and connect the dots that shape successful outcomes. While certain programs may necessitate the involvement of multiple TPMs with specialized focus areas, it remains crucial for you to immerse yourself in the entire software/product development lifecycle.

Embracing end-to-end ownership empowers you to have a holistic perspective on the program, enabling you to ensure timely and high-quality launches. By taking ownership, you dive deep into each stage of the process, fostering a profound understanding of the interconnectedness of tasks and the impact they have on the overall success of the endeavor. This comprehensive involvement not only enhances your professional experience, but also builds stronger relationships with cross-functional teams and stakeholders. It serves as a catalyst for improving your communication and stakeholder management skills as you become the trusted point person for inquiries and updates.

When you embrace end-to-end ownership, you hold yourself accountable for the success of the program. This accountability fosters a sense of confidence in your abilities and establishes you as the go-to person for insights, guidance, and decision-making. Colleagues and stakeholders recognize your commitment to the program's success, and they view you as a valuable resource capable of driving impactful outcomes.

However, it is important to acknowledge that managing multiple programs simultaneously may make it challenging to have end-to-end ownership of each one. Realistically, it may not be feasible to dedicate equal levels of involvement to all initiatives. To effectively navigate this complexity, it is recommended to allocate approximately 60-70% of your bandwidth to medium to large, complex, and cross-functional programs that demand maximum involvement and ownership. The remaining time can be split to provide execution support for small-medium sized programs, team rituals, or process improvement efforts.

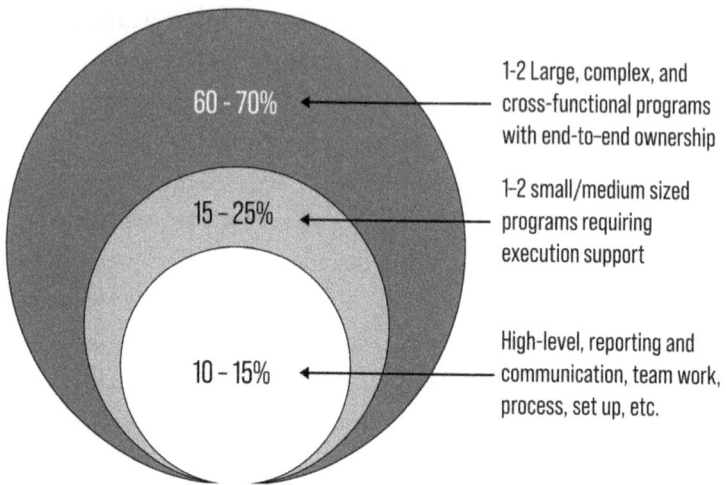

TPM Bandwidth Allocation

By prioritizing your level of engagement based on the program's size and complexity, you can ensure that you contribute meaningfully to its success.

PRO TIP: If you want to get to senior levels like Staff or Principal TPM, try to identify and drive at least one large complex and cross-functional program end-to-end. This will also give you a wide range of experience and help you become more versatile.

DEVELOP A HOLISTIC COMMUNICATION STRATEGY

Every program needs a communication plan in addition to an execution plan. You may also need a communication plan that is not limited to one program but covers multiple programs depending on your role and seniority. A strategic communication plan goes beyond sending regular status updates; it involves understanding your stakeholders, fostering collaboration, and ensuring everyone is aligned and informed about the program's progress and risks.

Doing so will not only keep the program on track, but will also help to build trust with the leadership team eventually, leading to achieving desired outcomes.

Consider the following to shape your communication strategy:

- **Format and Medium:** Format of the communication relates to information structure. This will be highly dependent on the medium of communication. Determine when it is appropriate to utilize meetings for discussions and when offline communication methods, such as email or program management tools, are more effective. Email communication will need to be formatted differently from a presentation, which will be different from a post or message in a group. Your audience is overloaded with information from every direction, so it is important for you as a TPM to make the communication worth their time. Your communication format should be able to quickly answer your audience's question, "What's in it for me?"

- **Desired Outcome:** Clearly define the purpose and desired outcome for you and your audience for each type of communication whether it's a meeting or email. For meetings, it can be brainstorming, decision-making, and for email it can be progress updates, requests for help, etc. The purpose of communication can change depending on the phase of a program. During the planning phase, your communication will focus on finalizing requirements or informing everyone about your roadmap, work breakdown structure, and timeline. During the execution phase, you may report weekly/biweekly status updates. Communication will also be needed if there are changes in scope or delays in the timeline.

- **Audience:** Identify your audience for each communication and tailor the information to their needs. Your audience can be the program core team, stakeholders from outside the organization, external vendors and/or senior leadership/ C-level execs. Consider what information is relevant and how it contributes to their understanding and engagement. Different people need different information, so you will need to adjust all aspects of your message based on the audience.

- **Level of Detail:** Think about how much time your audience has to read and absorb the information. Determine the right level of detail to keep stakeholders informed without overwhelming them with excessive information.

  Example: Senior leadership is most interested in knowing about key risks, mitigation strategy, and if they need to step in to alleviate the issue. On the other hand, a cross-functional team wants to know when dependencies will be resolved for them, so they can continue the required work.

- **Cadence:** Determine how often you want to communicate and whether you need multiple cadences depending on information type or audience.

  Example: You may do a weekly update for your core team but a biweekly update for other stakeholders.

- **Effort:** Be mindful of the time and effort required for communication activities. How much time do you want to spend putting together different types of communication? Strive for an optimal balance that allows for effective communication without becoming a burden on you or others involved.

Once you have a holistic communication strategy, you will need to start formulating the details that you want to get across. You are a connector between the technical and business world and therefore should be able to effectively communicate technical and business information to non-technical and technical audiences, respectively. You should make their communication crisp and insightful and you may need to adapt your communication style to the culture of the team/company.

Here's a sample format that can be used for multiple levels of audience, starting with a summary followed by more details.

SUMMARY

[Short 2-3 line summary/highlight - if someone only has 30 seconds, will they understand the purpose of this update]

WHO

[Who should read this? This does not need to be part of the communication but helpful to define the audience]

WHAT

[Description of your update/communication/ask]

WHY

[Context/rationale and importance of WHAT?]

WHEN

[Section to provide dates/timelines on specific activities]

WHERE/HOW

[As applicable]

DETAILS

[Long version as needed to provide deep dive for those who want to read more]

Sample Communication Template (Expanded template available on my website)

I also wanted to share seven communication best practices that you can utilize for both verbal (presentations, 1:1s) and written communication (emails, reports).

1. Offer concrete answers to questions and address the underlying "So what?" question that may come up.

2. Focus on the quality of your communication rather than quantity, so that you have the time to provide deeper insights.

3. Anticipate and address potential queries before they arise, so you can demonstrate foresight and a comprehensive grasp of the subject matter. This proactive approach instills confidence in stakeholders and reduces ambiguity, fostering smoother project execution.

4. Listen deeply and pay genuine attention to what others are saying, so you can gain a full understanding of their needs, concerns, and perspectives.

5. Ask specific questions to demonstrate a commitment to understanding and engaging with the subject matter at hand.

6. Provide clear and timely responses to help everyone stay on top of any issues and ensure there isn't any miscommunication.

7. Seek feedback on your communication to confirm that the message has been received and understood correctly and can make any adjustments if needed.

Example: If you see a couple of team members not aligned on a solution or idea, ask questions that will help them decide: "What is the core problem that each solution will solve?" or "How does each solution stand up to the future phases of the program?"

PRO TIP: For every communication, start by asking yourself "Who is my audience?" and "What do they want to know from me?"

MITIGATE RISKS PROACTIVELY

We discussed earlier in the book about how complexity and uncertainty lead to higher risk and why TPMs are desired. Therefore, risk management is one of the most crucial aspects of being a TPM. Minimizing potential issues results in smoother execution increasing the chances of success. By understanding and mitigating risks, you can anticipate challenges, make informed decisions, and maintain program timelines and quality. All the tools and methodologies utilized for program management are just a means to identify and reduce risks. Sharpening your critical thinking and decision-making skills will help develop a keen sense of risk. You can decrease the probability and the negative impact of risks, leading to better business outcomes. Risks are also the number one worry for most senior leaders and executives, so it is important to communicate risks and escalate if needed in a timely manner.

- **Identify and categorize risks:** Develop a comprehensive understanding of potential risks by conducting risk identification sessions with relevant stakeholders. You can categorize risks in many different ways such as using the matrix of what is known versus unknown or by different aspects of the program - technical risks, resource risks, external risks, etc. Do this as early as possible in the product development lifecycle. The planning phase is the best time to create a risk register.

- **Assess and prioritize risks:** Evaluate the potential impact and probability of identified risks to determine their significance. Use both qualitative and quantitative risk assessment techniques to prioritize risks that require immediate attention.

  Risk Level = Probability of failure x Impact of failure

- **Develop risk mitigation strategies:** Create proactive plans with detailed information like expected mitigation date, owner, etc. Your plan may involve implementing preventive measures, developing contingency plans, or allocating resources to minimize the impact of risks. You may also choose to escalate the risks for garnering higher level support. Escalation is not always a bad thing as long as it is backed by data and a contingency plan.

- **Establish risk monitoring and control mechanisms:** Continuously monitor identified risks throughout the product life cycle. Regularly review risk registers, update risk mitigation strategies, and establish feedback loops to ensure risks are effectively managed.

- **Communicate risks to stakeholders:** Maintain transparent communication channels to keep stakeholders informed about potential risks, impact, and mitigation strategies. Engage stakeholders in risk management decisions and let them know whether you are informing, escalating, or need help in a specific aspect of the mitigation plan. This will give them clarity on how the risk is progressing and how they need to engage.

PRO TIP: Risk reviews are a great way to share known risks. Most importantly, though, always share a mitigation plan and estimated risk mitigation date. If you don't have a date, provide a date for a date. This will build confidence with your stakeholders because they know you have a plan. It will also be an opportunity to get help from senior leaders if needed.

DEVELOP DATA-DRIVEN MINDSET

Embracing a data-driven mindset empowers you to make well-informed decisions, spot trends, and assess performance with precision. This approach not only elevates program outcomes and mitigates risks, but also streamlines the communication of valuable insights to stakeholders. Furthermore, data serves as a strong foundation for refining execution processes and gaining crucial support for new ideas. As AI is taking over the tech landscape, data becomes even more important for critical business outcomes. Your ability to harness the power of data can truly distinguish you from the field. If you are not already working with data, here are five steps to always take when managing any program.

- **Identify relevant metrics:** Determine the key performance indicators (KPIs) and metrics that align with program goals and objectives. Select metrics that provide actionable insights and reflect the program's success criteria.

- **Collect and analyze data:** Establish data collection mechanisms and processes to gather relevant data points. Leverage tools and techniques for data analysis, such as statistical methods, data visualization, and trend analysis, to derive meaningful insights.

- **Interpret and communicate insights:** Analyze and interpret the data to identify patterns, trends, and correlations. Translate the insights into actionable recommendations and communicate them effectively to stakeholders using data visualization techniques to enhance understanding.

- **Incorporate data into decision-making processes:** Integrate data-driven insights into the decision-making process such as launch go-no-go meetings. Ensure that data is considered alongside other relevant factors to make informed decisions that align with program and business objectives.

- **Continuously refine data collection and analysis:** Regularly review and refine data collection processes to ensure data accuracy and relevance. Incorporate feedback and lessons learned to improve the quality of data-driven decision making.

PRO TIP: Get technical with data to help you further. Learn SQL, Excel, Tableau, or other data visualization and reporting tools, so that you can analyze the data, irrespective of the data source or specific product metrics.

Data can also help influence stakeholders and senior leaders. Providing numbers can help assess a situation more objectively, resulting in a conversation that is focused on outcome rather than individual preferences.

EXPAND TECHNICAL SCOPE AND INFLUENCE

As you rise up the levels in TPM and manage bigger problems, your technical scope will expand as well. You will need to expand your technical understanding from one component or portion of the system to the entire technical stack depending on the type of program. This means that you may not know how every single component works within a system, but you understand how they interact with each other and can jump into one of the component areas if needed. Your understanding of the intricacies of the entire system will allow you to partner more effectively with other teams in the ecosystem. Besides your technical scope, your sphere of influence will also grow. You may also be driving programs that are cutting edge, which can bring their own set of complications. Think about broader implications of the technical system and inform the team of the same. There is more ambiguity, which requires you to be more vocal in technical discussions.

Developing Tech Responsibly

New technology undoubtedly enhances human life, but it can also yield unintended negative consequences, which demand our attention. For instance, while social media facilitates reconnections and support networks, it has also been a platform for harmful propaganda. Likewise, as AI gains prominence, it is important to be mindful of its future implications. As a TPM, you are in an influential position to ensure that new technology is being developed responsibly and that you consider the ethical dimensions of your work, including privacy, data security, and bias mitigation. It fosters trust with customers and stakeholders, mitigates negative impacts, and upholds ethical standards. It ensures that long-term success is prioritized over short-term hacks.

- **Incorporate ethics into program planning:** Integrate ethical considerations into the early stages of program planning. Identify potential ethical challenges and develop strategies to mitigate risks associated with privacy, data security, bias, and other ethical concerns.

- **Engage stakeholders in ethical discussions:** Foster open dialogues with stakeholders about the ethical implications of any initiative. Encourage discussions on responsible tech development and involve stakeholders in decision-making processes related to ethical considerations.

- **Stay updated with ethical guidelines and regulations:** Keep abreast of industry best practices, ethical guidelines, and regulatory frameworks related to responsible tech development. Incorporate these guidelines into processes and ensure compliance with relevant regulations.

- **Conduct ethical impact assessments:** Assess the potential social, economic, and environmental impacts of the technology being developed. Evaluate how the technology aligns with ethical principles and ensure that responsible practices are upheld throughout the product life cycle.

- **Promote responsible use of emerging technologies:** Advocate for the responsible and ethical use of emerging technologies within the organization and industry. Educate stakeholders about the potential risks and benefits of new technologies, and champion the adoption of responsible practices.

PRO TIP: Connect with other TPMs and industry leaders outside your organization to discuss how they are thinking about responsibility and ethics in developing new technology. Such cross-industry collaboration will benefit everyone.

MANAGE CHANGE AND ADAPT QUICKLY

Despite our best efforts to minimize changes during program execution, change remains an inevitable part of the process. Agile methodology was developed precisely to help manage change more smoothly. Changes may include altering product requirements, team members leaving, or tools changing. Effective change management ensures that these transitions happen smoothly, reduces resistance, and increases the likelihood of success. Embracing iteration and adaptation helps navigate unforeseen challenges while keeping the program moving forward. To simplify the process, here are five tips for managing change effectively.

- **Create a change management plan:** Develop a comprehensive plan that outlines the objectives, scope, and stakeholders involved in the change initiative. Clearly define roles and responsibilities and establish a timeline for implementing the changes.

- **Communicate change quickly:** Use clear and consistent communication channels to inform stakeholders about the upcoming changes. Explain the rationale behind the changes, highlight the benefits, and address concerns and questions to foster acceptance and buy-in.

- **Manage stakeholders' expectations:** Engage with stakeholders to understand their expectations, concerns, and needs. Tailor communication and change strategies to address stakeholders' unique requirements and ensure their support throughout the change process.

- **Address resistance to change:** Proactively identify potential sources of resistance and develop strategies to address them. Engage stakeholders in the change process, provide training and support, and highlight the positive impact of the changes to alleviate resistance.

- **Foster adaptability and resilience:** Encourage a culture of adaptability and resilience within the team. Foster an environment where team members feel empowered to embrace change, experiment with new ideas, and learn from setbacks.

* * *

PRO TIP: Change can be unnerving so be mindful of how you handle change, especially in front of your team. Develop a curiosity mindset whenever dealing with change. Asking questions will help you get more information, which will be useful for handling the change quickly.

CHAPTER 18

PLANNING YOUR CAREER PATH

*A career path is rarely a path at all. A more interesting life
is usually a more crooked, winding path of missteps, luck
and vigorous work. It is almost always a clumsy balance
between the things you try to make happen and the things
that happen to you.*

– Tom Freston

Whether you are an aspiring TPM or have some experience, you
will soon start thinking about growing into the role and exploring
career options. Because the role has matured over time, the
possibilities within the TPM career spectrum have also evolved
significantly. TPMs have an array of career options that they can
choose based on personal interests and areas of strength. TPMs
also have many transferable skills that open up several career paths
as it is unnecessary to stick to one path.

ASCENDING AS IC OR VENTURING INTO MANAGEMENT

TPMs can grow as an IC, progressing through different levels
(ranging from Associate (L3) to Principal (L8) level). This pathway
allows TPMs to dig deeper into the complexities of technical
program management, refining their skills as they contribute to
larger initiatives. TPMs also have the choice to shift to a management

track, taking on the responsibility of managing other TPMs. This managerial role entails building and managing their teams, aligning strategies, and fostering collaborative environments. TPMs on the management track can advance from a level one manager all the way to VP. Realistically, in most companies, the management track for TPMs ends at the Director level. Only the big tech companies have VP-level positions within the TPM function, but they are less and far between. However, that doesn't mean the end of your career. Today, there are many more career tracks available that leverage the core skills of a TPM or TPM leader. You can always move laterally into a different function or role at those senior levels.

HYBRID APPROACH: BALANCING MANAGEMENT AND HANDS-ON CONTRIBUTION

Explore the hybrid option, which, similar to a Tech Lead Manager, is gaining popularity. Here, a TPM lead combines hands-on program management with mentoring or managing a smaller group of 1-3 TPMs. This option is best for people who want to retain direct involvement in programs while also engaging in limited people management responsibilities. It demands a delicate equilibrium between technical acumen, interpersonal skills, and strategic oversight.

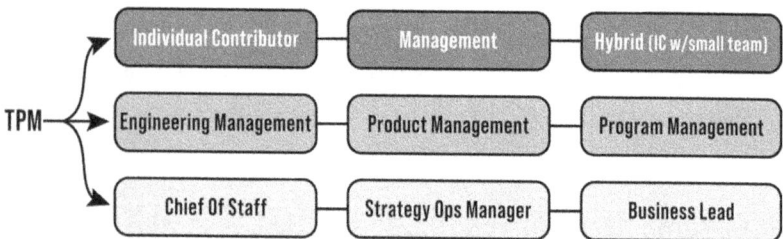

TPM Career Path Options

TRANSITIONING TO ENGINEERING OR PRODUCT MANAGEMENT

Many TPMs often move into engineering or product management roles for various reasons. Some are drawn to the technical aspects, while others seek higher pay or increased visibility. These roles often offer a clear career path that may align more closely with your goals. However, keep in mind that making these transitions may require gaining new skills related to those roles. While it's a good choice, don't use the TPM role as just a stepping stone because it can create confusion for stakeholders if you are acting like an EM or PM. It's important to continue honing your TPM skills because they remain valuable, even in other roles.

VENTURING INTO BUSINESS-FOCUSED ROLES

Another intriguing trend sees TPMs moving into business-focused roles that encompass strategic perspectives. Positions such as Chief of Staff, Strategy Operations, and Business Lead leverage the strong program management skill set of TPMs while directing it towards broader business strategy and operations. These roles require a heightened understanding of an organization's strategic goals and a keen awareness of its operational intricacies. In fact, TPMs have a great foundation for these roles if their interests are not purely technical. They can choose any area and advance on that ladder, opening up bigger possibilities.

The TPM career path doesn't have to adhere to a strict linear progression. In fact, a hallmark of TPM career growth lies in the ability to traverse back and forth between different TPM archetypes and even between different roles. This lateral movement enriches one's experience, providing a multifaceted view of program management and a well-rounded professional profile.

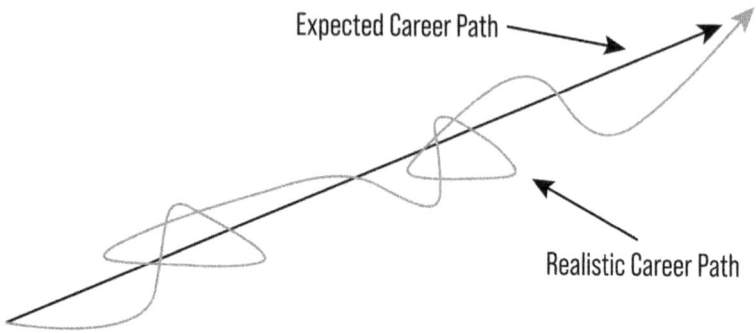

Expected Career Path

Realistic Career Path

Your Career Path Is Not a Straight Line!

PATH TO C-SUITE

Your TPM Career can lead to great heights given your sound foundation. While it is easy to think that an engineer can become a CTO and a product manager can become a CPO, I want you to know that there's nothing stopping a TPM from reaching the C-level position. I encourage you to tap into your strengths, experiment with different roles, and choose the trajectory that best works for you. Today, pivoting careers is common, so if you are hitting the ceiling on the TPM track, seek other opportunities that may provide the challenge and a path to C-Suite. A TPM has many transferable skills (program execution, stakeholder management, operational excellence, communication, leadership, etc.) for positions like COO, General Manager (GM) of a business unit or even a CEO.

Anna Folley is someone who has broken into the C-Suite with her exceptional skills and experience. She moved into the CIO (Chief Information Officer) role at a ride-hailing startup after establishing their Technical Program Management function, leading strategic, cross-functional programs within the Technology group. She was instrumental in getting the technical infrastructure ready for IPO, building a world-class technical onboarding curriculum, and many more significant endeavors that drove the company's growth and enabled the organization to scale, quickly and effectively.

* * *

PRO TIP: Try out the hybrid option if you are unsure whether the management track is right for you. You will get practical experience managing people and understanding what it takes for a people manager to succeed.

TPM SPOTLIGHT

BEN GAUTHIER, SENIOR MANAGER, TPM

Ben pursued an undergraduate degree in physics and mathematics, but the prospect of being confined to a lab didn't resonate with him. This led him to explore diverse paths, starting with a Wall Street internship. Eventually, he found his way into business rotations at a big corporation, where he worked in finance, mergers and acquisitions, and briefly worked at their retail counter! His blend of technical and financial skills led him to a software engineering role within the finance domain of a big tech company. He began his TPM career with conversational AI and eventually joined a startup to work on autonomous vehicles. He has since then transitioned into a managerial role.

What motivated you to move into TPM?

Ben: The desire to combine my technical and soft skills drove me towards TPM. I had no idea the TPM role even existed for the first years of my career until my then-manager introduced me to it. This was a dream come true - for the first time I felt there was a role that gave that perfect balance I was looking for. Soft skills are crucial for a TPM who is enabling effective communication about complex matters and navigating challenging situations. At the same time, without deep technical skills it's near-impossible to properly execute on a complex technology and earn trust with engineers.

What surprised you as you navigated the TPM role?

Ben: A striking realization was that everyone, including me, is essentially "making things up as we go." Transparency and quick decision-making became key. Navigating ambiguity and conveying complex information were challenges in my early days. It dawned on me that nobody possesses a comprehensive understanding of what's transpiring at all organizational levels. Thus, I focused on gathering information and adapting or failing quickly if needed.

What do you like about being a TPM?

Ben: The dynamic blend of technical and soft skills is what I enjoy most. This role empowers me to manage without formally being a people manager. I derive satisfaction from assisting and unblocking team members, as well as collaborating with various stakeholders.

What do you dislike about the TPM role?

Ben: A drawback is that regardless of what goes wrong in a program, it can be tied to the TPM. Maybe the timeline was set too aggressively, the risk wasn't escalated fast enough, or an engineer found a hidden bug in their code the day before launch; no matter what the root cause is or who "messed up," the TPM is the one who has to stand up and explain that last minute delay to leadership. This aspect, coupled with the constant pressure to deliver, can impact one's mental well-being. It's important to choose battles wisely since not every situation can be fixed or foreseen.

What do you think is the best way to leverage TPMs in an organization?

Ben: TPMs are valuable but come with a cost - a company often needs to choose whether to hire a TPM or an extra engineer. Not all organizations even require them. TPMs excel in program launches, aligning cross-functional tech components, making critical technical decisions, incident management, and ensuring hardware-software alignment across different disciplines. So TPMs should be leveraged in programs where these skills will be highly valuable.

How do you scale yourself when managing large, company-wide programs?

Ben: I realized early on that there's a limit to what I can handle, which led to setting boundaries. Scaling involves two key components: program management style and communication. It's impractical to micromanage tasks beyond an entry-level TPM. It is important to understand that trying to "do everything" means you will not be able to scale to drive large programs. Moreover, there is no perfect way to manage multiple teams and programs that work in every situation, especially not using Gantt charts.

I focus on starting with the minimum and progressively building up. Clear communication, defining ownership, setting expectations, delegating, and using templates is instrumental. In fact, I did not personally file a single Jira ticket when handling one of our critical company-wide programs - my ROI instead came from setting process and structure and empowering my tech leads.

Can you share more about what actions you took to manage one of your largest programs?

Ben: This was a massive program that at its largest point involved over a hundred engineers and six engineering VPs. The requirements were highly complex and safety was of the utmost importance. This meant there were hundreds of details to be taken care of. To manage this scale, I established strong partnerships with technical leads in each organization. We formed "strike teams" which were smaller groups of people that collaborated intensely to achieve milestones. I focused on fundamental questions: "What's the Definition of Done (DoD)?", "Who do we need to make this work?" and "What is the planned work?" These questions helped set clear expectations.

Then, I empowered each strike team with templates and ownership, ensuring they had the authority to independently make their own decisions wherever possible. I consolidated reporting and communication to provide centralized updates to leadership. Additionally, I didn't shy away from seeking assistance and delegating tasks. This allowed me to manage this intricate program efficiently and support my direct reports while still maintaining a healthy work-life balance.

This experience highlighted that with proper planning, effective delegation, and streamlined communication, even intricate programs can be managed successfully. The approach I adopted here was praised by senior leadership and recommended as a model to be adopted throughout the program management team.

What motivated you to move into management?

Ben: In a meeting with my manager at the time, I remember I made the deliberate decision to remain at Level 6 and become a manager instead of targeting a promotion to the next level. My initial motivation was the desire for personal growth and the chance to tackle a whole new challenge that pulled on my years of TPM experience. However, my interest in caring for people and tackling human-related challenges rapidly became my key drive. I realized that helping and growing others had been one of the most intrinsically rewarding parts of my career. I wanted to help people succeed faster while leveraging their strengths. Best of all, the manager role gave me the opportunity to leverage my strengths in communication, emotional intelligence, and executive presence.

Honestly, being a manager also gives me a lot of visibility despite the stress. There is almost 20% additional work than being an IC at the same level, but the opportunity for impact is higher.

What was different for you being an IC vs. manager?

Ben: As a manager, I'm more concerned about my team's well-being and performance, almost resembling the role of a parent. This is in contrast to when I was an IC, where my primary focus was on the success of my own work versus that of my peers.

What advice would you give other TPMs or aspiring TPMs?

Ben: Don't overlook the importance of soft skills. Becoming a good TPM is relatively straightforward, but scaling to a great TPM without strong soft skills is a challenge. Confidence is vital—know when to speak up and have the ability to coach your team. Strive to display confidence while also learning from others.

PUT IT INTO PRACTICE

✓ Download the 30-60-90 Day Goals Template from my website and use it for every new job and role. Check the Resources section at the end of the book.

✓ Analyze the expectations for you at the current and next level (as aligned within your company) and categorize into Green - Doing consistently; Yellow - Doing inconsistently; and Red - Not doing at all or not doing well. This will help you with promotion readiness.

✓ Make a list of all of your tasks and categorize them two ways - Strategic versus Tactical and Urgent versus Important. Then prioritize the DO list while delegating or deprecating others.

Part 4

FINDING TRUE SUCCESS

Success in life comes to those who simply refuse to give up;
individuals with vision so strong that obstacles,
failure and loss only act as teachings.

– Silken Laumann

CHAPTER 19

DEVELOPING
EMOTIONAL INTELLIGENCE

Emotional intelligence, more than any other factor, more than I.Q. or expertise, accounts for 85% to 90% of success at work… I.Q. is a threshold competence. You need it, but it doesn't make you a star. Emotional intelligence can.

– Warren G. Bennis

Jane was a highly competent TPM on my team who excelled at getting things done and delivering results. She had the ability to take on high-risk programs and deliver them on time, meeting all parameters of success. Not only was she technically strong, but she also possessed excellent execution skills that helped the team achieve business goals and make a significant impact. However, she often came across as adversarial and demanding, insisting others do things her way. Her abrasive language made people feel disrespected and alienated. Regrettably, she mistook this behavior for assertiveness and leadership that, unfortunately, led to unintended consequences, leaving her colleagues unhappy and unwilling to work with her. The EM and PM escalated the situation to me and I observed firsthand the tense and strained atmosphere during meetings and how her lack of understanding of other people resulted in losing trust and credibility with the team in spite of great program management skills. Her impact was diminished because her behavior was deemed hostile. As her manager, I needed to step in to do damage control and regain our partners' trust.

Your impact is not just about what results you deliver, but also about how you deliver them. What behaviors do you demonstrate and how do your actions lead other people together towards a common goal?

IMPACT = RESULTS + BEHAVIOR

In today's rapidly evolving and highly dynamic tech workplace, technical and program management expertise alone is no longer sufficient for success. Multiple teams are involved in bringing a product or service to market. TPMs are often working with globally distributed teams and the need to possess exceptional interpersonal skills becomes increasingly critical. This is where emotional intelligence (EQ) steps in as a vital factor that sets individuals apart in their careers.

Emotional intelligence refers to the ability to recognize, understand, and manage one's own emotions, as well as empathize with and navigate the emotions of others. It encompasses self-awareness, self-regulation, social awareness, and relationship management. While technical skills are undoubtedly important in the tech industry, EQ plays a significant role in determining an individual's effectiveness in collaboration, leadership, and overall professional growth. In people-oriented roles such as TPM, having a high level of EQ makes the job easier. When people feel respected and heard, they are more inclined to collaborate and support you. By being aware of and in control of your own emotions while understanding the emotions of others, you can develop deeper relationships with others. This not only builds social capital, but also enhances other crucial skills such as communication and thought leadership.

Today, teams are often diverse and dispersed, and rapid innovation drives constant change: EQ provides a competitive edge. It allows you to establish strong connections with colleagues, build trust, and effectively communicate ideas and solutions. EQ empowers you to navigate challenging situations, resolve conflicts, and inspire others to collaborate and achieve shared goals. As technology continues to reshape industries, the human element remains essential. Emotional intelligence helps bridge the gap between technology and human interaction. It enables you to understand the needs, desires, and motivations of end users, ensuring that technological solutions are not only functional but also resonate with the intended audience.

By honing your EQ skills, you can unlock the full potential of your technical abilities, foster innovation, and create a workplace culture that thrives on collaboration, empathy, and mutual respect.

EQ AND EMPATHY

Empathy plays a vital role in EQ. When you genuinely understand the feelings of others and can see situations from their perspective, it becomes easier to find win-win solutions. Empathy facilitates conflict management, enabling you to identify the true root causes of problems. For example, Jane could have greatly benefited from understanding the viewpoints of other team members, acknowledging their competing priorities, and recognizing that hers was not the only important initiative.

Awareness of your own emotions, actions, and their impact on others allows you to gradually uncover your own biases. By approaching your own feelings with curiosity and asking questions of others to learn more, you can deepen your understanding of yourself and others. Additionally, seeking feedback from your manager, peers, stakeholders, and others is essential. Feedback serves as a valuable tool for personal growth. When receiving feedback, listen with an open mind and avoid becoming defensive.

Though it can be challenging to receive critical feedback, approach it with curiosity, asking questions to gain a deeper understanding.

Many companies often index empathy signals when making hiring decisions. Adopting this mindset fosters personal growth, enabling you to achieve greater things for yourself.

EQ AND EFFECTIVE COMMUNICATION

As you fine tune your EQ, you begin to understand your own communication style. You understand how to cater to your audience's needs. Communication becomes truly effective when our audience comprehends the content and achieves complete clarity. The first step towards impactful communication, whether written or verbal, is to understand your audience through the power of EQ and empathy.

Are you engaging with C-level executives, senior leaders, managers, stakeholders, or peers? Each group possesses distinct interests and information requirements. Rather than creating numerous reports, you can adapt by structuring your communication to provide an overview while highlighting specific details that some individuals may find valuable. By addressing our audience's primary concerns, you enhance clarity and minimize confusion.

In today's age of information overload and fleeting attention spans, developing your EQ can help capture and retain your audience's attention. People tend to scan emails within seconds, making it essential to quickly engage them. Being concise and sharing relevant information enhances clarity in your communication. You must clearly articulate what you are sharing, why you are sharing it, and how you are actively solving problems. EQ helps you recognize that managers and leaders rely on your information to then communicate up the chain. The more they are informed, the less anxious they become, fostering greater confidence in your abilities.

For instance, highlighting risks is a good thing and you have to understand what questions your audience may have and what they care about when they hear the news. If you highlight a risk, it is crucial to provide information on when the risk will be mitigated and explain the steps being taken. Even if a specific date is not available, providing a timeline for an update demonstrates proactive management. If a metric is trending downward and you only report it without further explanation, the information becomes useless. Instead, provide additional details about the implications, potential impact, risks, and the steps you or your team is taking to address the situation. The senior leaders will have more confidence in you and they can be clear on what they need to do based on your update.

I used to include a riddle or joke at the end of some of my reports to finish on a lighthearted note and see if I could gauge audience engagement and identify those who read the report until the end. You don't have to do the same (and you have to be mindful of culture), but you should still find ways to measure the effectiveness of your communication.

EQ AND NURTURING RELATIONSHIPS

Building strong and trusting relationships is a prerequisite for influencing people and aligning teams. When there is no prior relationship, requests can come across as transactional, hindering desired outcomes. Non-personal communication poses a barrier to achieving mutual goals. To overcome this, you must connect with people on a deeper level and genuinely understand them. Initiating relationships without expectations or hidden agendas opens the possibility for lasting friendships to emerge from honest work connections. For that, you need to identify the right folks and connect with them outside the proceedings of specific projects or programs that you may be involved in together. You can identify who you want to meet, how often, and the best way to develop a relationship based on their circle of influence:

- **Circle 1:** Senior stakeholders who are 2-3 levels above you like senior managers, directors, etc. These are folks that are interested in the overall progress of the organization and are often influential.

- **Circle 2:** Partners and cross-functional peer managers like engineering manager, product manager, etc. You may work with these folks often in the context of a program, and you all may share similar goals.

- **Circle 3:** Members of engineering, TPM, or other teams who may need to contribute to a program or are part of your broader community. You may not connect with them every week for a program, but you want to have a relationship to learn, share, and help each other if needed.

Building strong relationships is one of the crucial aspects of growing in your career and it requires you to develop strong emotional intelligence. Doing so will help get a better understanding of others, which will serve as the foundation for effective communication, successful project execution, and fostering a positive work environment. Aim to be authentic and reliable in your interactions with your stakeholders so you establish trust quickly. Share information, insights, and updates to foster a sense of inclusion and help others understand your thought process. Your transparency helps create an environment where individuals feel comfortable sharing their own ideas and concerns. This leads to better collaboration and problem-solving. When you understand the perspectives and challenges of others, it shows that you genuinely care about their well-being and success. When people feel heard and respected, they are more likely to align with your intentions and be receptive to your ideas and feedback. Your actions and words can have a big impact on others, so consider how your behavior might be perceived and how it aligns with your intended message. Strive to be consistent in your approach, treating everyone with fairness, respect, and dignity. It is all too easy to throw someone under the bus accidentally when trying to meet goals.

Example: While reporting "Task/owner XX is behind and Engineer Z owns it" can be an accurate description of the current situation, it can come across as pointing fingers. A better way could be to bring this up with the owner and share their point of view in the update: "Task XX is delayed due to conflicting priorities" or "Task XX is pushed back by three weeks due to ABC reasons, but Engineer Z has put a mitigation plan together."

* * *

PRO TIP: Check out the book "Positive Intelligence" by Shirzad Chamine, which is a great resource for understanding one's own mindset, its impact on our performance, and building empathy for ourselves and others. The book also dives into how to build a positive mindset, which helps navigate tricky and high-pressure situations more effectively.

CHAPTER 20

DEMONSTRATING THOUGHT LEADERSHIP

Thought leaders know how to take an idea and put it into action.

– Dillon Kivo

In the first month after joining a big tech company, I set up an introductory meeting with my organization's director. A few minutes into the meeting, he posed an unexpected question: "Are you considered a thought leader by your team?" I don't remember what I said, but I remember feeling stupid and confused. What does thought leadership truly mean? Is there something I am not doing? As time went on, I realized that thought leadership is closely tied to influencing others. It involves bringing forth your unique perspectives, ideas, and opinions to drive meaningful conversations and garner recognition for your contributions.

As you progress in your career and take on the responsibility of managing big programs, becoming a thought leader is essential. Thought leadership also helps in creating a brand that distinguishes you for your ideas.

BE CURIOUS AND ASK QUESTIONS

Curiosity is a key trait of effective leaders. They are always eager to expand their knowledge on various subjects. Every year Barack Obama or Bill Gates share their list of book recommendations, which goes to show how they continue to find inspiration through curiosity. Learning doesn't have to be limited to formal training, you can absorb insights from diverse sources like podcasts, books, audiobooks, web articles, TED Talks, and conferences. The more you engage with these, the more your brain retains valuable information. Over time, this knowledge can spark new connections and inspire original frameworks and ideas.

Another crucial aspect of curiosity is asking questions—something we naturally excel at as children. However, as we grow into adulthood, we sometimes shy away from asking questions, fearing it might be perceived as ignorance or incompetence. This can lead to one making assumptions and jumping to conclusions in order to provide a smart answer. This misconception can stifle curiosity and hinder personal and professional growth. Cultivate the skill of asking thoughtful questions throughout your life.

As you progress in your career, people will often look to you for answers. Yet, it's equally important to seek clarity by asking questions in many situations. While sharing your opinions is vital for developing thought leadership, curiosity has to come first. In roles like TPMs, asking probing questions—even if you may already have some answers—can uncover valuable insights. Being the devil's advocate at times and asking more questions can lead to unexpected, new information. Thoughtful and well-crafted questions challenge assumptions, encourage reflection, and uncover hidden opportunities. By asking probing questions, you foster a culture of intellectual curiosity and facilitate the exchange of diverse viewpoints.

LISTEN AT LEVEL 2 AND 3

Have you found yourself in a meeting, listening to someone speak, yet your mind is already racing, formulating how you'll respond or what you want to say next? This also occurs in smaller group or one-on-one conversations too. In these moments, your focus is split as you try to craft a response or counterargument while the other person is speaking. In other words, you are listening passively. This results in disengagement and loss of connection as your mind has already drifted. You might catch the first few sentences but risk missing important details. To offer considerate responses, it's crucial to embrace what is called active listening where you are paying full attention to the speaker.

Listening happens at three different levels:

- **Level 1:** Internal Listening is listening to yourself, or your own thoughts or agenda. At this level you are listening only to respond or impress. You are probably thinking, "What does this thing have to do with me?" or "Do I have anything interesting to say?" Judgment of self and others also occurs at this level.

- **Level 2:** Focused Listening is listening intently to another person without getting distracted. In other words, you are listening to learn, you are observing their tone, interested in their story, and asking, "How do I know more?"

- **Level 3:** Global Listening is listening to others in the context of the entire surrounding. At this level, you are listening to understand deeply by observing body language, the inflections, pauses and hesitations. You are using your intuition to understand what is not being said and wondering how you can best support the speaker.

Understanding these three levels of listening can help us improve our listening capabilities. Active listening happens at Level 2 and Level 3, which helps you engage in deep conversations.

Active listening requires giving your full attention to the speaker, seeking to understand their perspective, and empathizing with their experiences. Your own agenda is set aside at that moment. Engage in genuine dialogue, encouraging others to share their thoughts and forward collaboration. Active listening also enables you to be curious and ask powerful questions that can stimulate critical thinking and lead to new insights.

> **PRO TIP:** Avoid multitasking during meetings and focus on listening. Build awareness on when your mind drifts away and go into solution mode to bring it back to the present. Build this practice and slowly you will find that you are becoming more focused. Sometimes, focusing on the speaker, making eye contact, etc. can help you retain focus.

VOCALIZE YOUR IDEAS AND OPINIONS

How often have you been in meetings and had a great idea but didn't share it, and then someone else shared the exact same idea? Showcasing thought leadership begins with the courage to share your ideas and opinions. By bringing your unique perspectives to the table, you contribute to the richness of discussions and enable creative thinking. Whether it's during team meetings, brainstorming sessions, or informal conversations, seize the opportunity to express your thoughts in a clear and concise manner. Present your ideas with confidence, supported by logical reasoning and relevant examples. Take the time to learn about the domain or technical details if needed and vet them with subject matter experts. This will ensure that your ideas are coming from a place of knowledge and understanding of the subject. By initiating thought-provoking conversations, you capture the attention of others. When your ideas resonate with people, they will eagerly anticipate more from you and may even reach out to collaborate with you. Find the right opportunities and mediums to express your thoughts, whether it be through meetings, one-on-one discussions, or written communication. Start with a medium that you feel most comfortable with and gradually expand your reach.

CHALLENGE THE STATUS QUO

When I first joined a big startup, I observed that there were a lot of inefficiencies in our weekly execution reporting to the SVP. The weekly meeting involved 20+ people including two separate department VPs and their Directors. Some of the topics being discussed could have been easily communicated offline. Besides, at least 10 different teams across those two organizations had their own meetings to prepare a highly comprehensive report that even included engineering code check-in information. I connected with two VPs to understand what would be most helpful to them going into the meeting with the SVP. One of the VPs was really set in his way and did not want to change anything about the reports or the meetings. They wanted to see all the information at the most granular level, which perplexed me. Eventually, through a lot of probing questions, I understood they were trying to assess the risks on their own because nobody was giving them the needed information. That gave me good insight and I reached out to a few managers and directors to get their feedback and identify any additional pain points. I came up with alternatives and discussed with the managers who were spending a lot of time creating the reports. Once I got alignment from the managers and VPs, I moved to address the big meeting between two engineering departments. It took a few additional iterations. Culturally, they had been running that meeting the same way and no one really thought it needed a change. However, as the changes led to higher quality of discussions, people saw the benefits.

To grow as a thought leader, you need to have the willingness to challenge the status quo if you have evidence that it is preventing the team or organization from maximizing their potential. This means you may question established norms, practices, and assumptions within your organization if you think they are not serving the current state of affairs. By challenging the status quo, you open doors to innovation, improvement, and growth. However, as you can see from my experience, that change takes time and you have to navigate the situation carefully.

Challenging established ways of doing things can feel daunting, especially if you fear the potential consequences. Here's how you can drive positive outcomes for your teams:

Remember the goal: Recognize that questioning the status quo is not about being rebellious or disruptive. Instead, it is a means of driving progress and uncovering new opportunities. In order to reach your goal, demonstrate a commitment to continuous improvement and a willingness to push boundaries for the betterment of the organization. This means that sometimes you have to compromise if you learn new information that no longer holds the initial argument.

Gather evidence and prepare your case: If you are confident that something needs to change, gather relevant data and evidence to support your argument. This includes research, facts, and examples from other organizations or industries that have embraced change successfully. Develop a well-rounded case that clearly outlines the potential benefits and addresses any potential concerns or objections. You are more likely to get support when your stakeholders see the effort you have put in and the data itself.

Find allies and build support: Seek like-minded individuals who share your concerns or desire for change. Engage in discussions, build relationships, and form alliances with colleagues who are open to challenging the status quo. Get their input and build a plan together. You are also more likely to get buy-in from senior leadership when you come across as one unit.

Start with small, incremental changes: If the fear of speaking up is overwhelming, begin by challenging the status quo in small, incremental ways. Identify areas where improvements can be made without disrupting the entire organization. By implementing small changes and demonstrating positive outcomes, you build credibility and confidence in your ability to effect change.

Choose the right time and place: Timing is essential to see desired results, especially where big changes are concerned. Select appropriate moments, such as team meetings, brainstorming sessions, or strategy discussions, where new ideas and perspectives are encouraged. You may also need to think if the organization is ready for your ideas or if there's any competing activities that may become an obstacle. If your ideas are "ahead of time," take it slow and get a feel for the appetite for change. By choosing the right time and place, you increase the likelihood of your ideas being accepted.

Communicate with empathy and respect: When presenting your case, ensure that your communication style is empathetic and respectful. Make sure you are not coming across as boastful or rigid. There have been numerous occasions where individuals say, "this is how it was done at my previous company or team." Keep in mind that the new team is different from your old one, so you want to frame your ideas in a way that highlights the potential benefits for this team and the individuals involved.

Seek feedback and learn from resistance: Change is one of the hardest things for any individual and even more difficult as the team/organization size grows. Be prepared to be challenged yourself and expect that there may be resistance and pushback. Instead of viewing it as a deterrent, see it as an opportunity for growth. Being open to feedback and different viewpoints will foster constructive dialogue rather than a confrontational atmosphere. Listen attentively and try to understand the concerns and how the idea impacts others. This will enable you to refine your approach, address objections, and learn quickly from differing viewpoints.

Making change is an ongoing journey, so stay curious and remain open to iterating on the change.

Communicate success: When your ideas lead to positive outcomes and change, take the time to share the impact of those initiatives with key stakeholders, highlighting how challenging the status quo has improved processes, increased efficiency, or driven innovation. By showcasing tangible results, you build a track record of success that can inspire others to embrace change. Aim to create a culture of innovation, adaptability, and growth within your organization.

Remember, change starts with a single voice, and as a leader, your willingness to challenge established norms can pave the way for meaningful transformation.

BUILD YOUR PERSONAL BRAND

Building your personal brand in this age of social media and technology is required more than ever before. A strong personal brand establishes your reputation as a subject matter expert and someone who consistently delivers value. Building your brand requires consistency, authenticity, and a commitment to continuous growth. Share your expertise through various channels, such as writing articles, presenting at conferences, or participating in industry forums. Actively engage in professional networks and communities to connect with like-minded individuals and exchange insights. Be intentional about nurturing relationships and demonstrating your expertise in a way that aligns with your personal values and professional goals. As your reputation grows, you become recognized as a trusted authority in your field, enabling you to attract opportunities and influence others on a larger scale.

* * *

Thought leadership not only contributes to the success of the programs you manage and the overall growth of your organization, it also shapes your professional reputation. Your ideas can lead to invocation and position you as a respected leader, able to influence others, drive meaningful change, and make a lasting impact in your field. Steve Jobs is a classic example of someone who disrupted the tech world with his innovative thinking. We were all happy with a Walkman/Discman, but the invention of the iPod and then the iPhone revolutionized the industry.

PRO TIP: Get into the habit of writing on a regular basis - preferably daily. You don't have to restrict yourself to write on work related topics or even publish what you write. Writing on any topic will help you develop your thought process and improve your delivery of ideas.

CHAPTER 21

GROWING AS A LEADER

Good leaders build products. Great leaders build cultures.
Good leaders deliver results. Great leaders develop people.
Good leaders have vision. Great leaders have values.
Good leaders are role models at work. Great leaders are
role models in life.

– Adam Grant

Back in my early career, like many others I had the misconception that leadership and management were pretty much the same thing. I believed that being a leader is only needed once you have folks reporting to you. As time went by, I continued working closely with people to achieve the results I wanted. As I climbed the career ladder, I began to realize that I'd actually been using some leadership skills all along but that there were more I needed to develop.

Being a leader in the professional world can be quite a challenge. There's an abundance of resources out there–books, podcasts, and training programs–all geared toward teaching you the ropes of effective leadership. But when you think about it, how many truly outstanding leaders can you actually name?

The term "manager" is tucked into "TPM," which essentially meant my role required me not only to manage programs and resources but, more importantly, to lead people–or more precisely–

to guide their output. To thrive in a role that involves interfacing with people at all levels of an organization, personal growth as a leader is absolutely essential. If you put your focus on becoming a leader before targeting a people manager position, it can lead to greater impact and success in the long run.

This chapter cannot cover every aspect of leadership. I will be focusing on a few key leadership skills that are essential for TPMs to elevate themselves and be seen as influential leaders by their organization.

TAKE OWNERSHIP

In my very first job, there was a fellow engineer named Andy who joined our team a few months after I did. Over the next four years, he experienced a rapid ascent, going from being a mid-level engineer to becoming a manager, and eventually a director. As I looked back on his impressive journey, I realized that it was built on tackling high-priority, complex projects, which weren't flashy and glamorous. These projects demanded serious effort, which many people shied away from because they weren't the "shiny new object." But Andy was different. He stepped up, volunteered to take on these challenges, and successfully resolved them, making a lasting impression on our management.

There's a motto prominently displayed in the new-hire orientation room of a big tech company that boldly states, "Nothing at this company is someone else's problem." The message is clear: if you're willing to tackle any important problem you encounter, your efforts won't go unnoticed. Whether it's a problem that's slipped under the radar or one that's been brushed aside because it's tricky and lacks glamor, don't hesitate to make it your own if it is still a priority for your organization's leader. These challenges offer a unique opportunity to rise to the occasion beyond your assigned duties. Find innovative and resourceful solutions to long-standing issues that have been plaguing your team or organization. By proactively taking on these tough problems, you have a chance to showcase your leadership abilities, earn the trust of management, and make a significant impact on the organization.

Here are nine key strategies that you can adopt to help you solve organizational problems that no one else wants to address.

Bias for Action

When confronted with organizational problems, take the lead instead of hesitating or waiting for someone else to bring it up or assign it to you. Be proactive and take the initiative to tackle the problem head-on. By doing so, you demonstrate your determination, agility, and willingness to confront challenges. You are also developing a sense of urgency to tackle the challenges head on which is needed for all leaders and results in impactful business outcomes.

Understand the Underlying Problem(s)

To gain a comprehensive understanding of the problem, do your research. Engage with different stakeholders from various functions and levels within the organization. Each person can provide valuable perspective, enabling you to develop a complete picture of the problem at hand. Your critical thinking power may uncover the root cause, which may not be visible on the surface. This will help you create the right solution that solves the underlying problem rather than fixing a symptom of it.

Ask Specific Questions

During your conversations with stakeholders, ask specific questions to gather as much information as possible. Inquire about their specific issue with the large problem, obstacles their team is facing, or how a solution would benefit them.

Connect the Dots and Find a Pattern

After gathering information from all stakeholders, review your notes and identify commonalities and patterns. Look for recurring themes or issues that emerge across different functions or teams.

Understanding the interconnectedness of problems will help you formulate an effective solution.

Prioritize One Key Problem

Once you are done with your research, you may find that there are multiple issues that could be solved. Attempting to solve all problems simultaneously is impractical and inefficient. Instead, focus on prioritizing one or two key issues that are critical to the business and can have a significant impact. Select a problem that will challenge you and have the biggest ROI for your team. Align with your manager or a key stakeholder to ensure that you have the support to solve the problem, especially if you are taking on something in addition to your other programs. Ensure that you are still able to meet any existing goals.

Collaborate on Solutions

Don't tackle these problems alone. Recognize that you may be working on this alongside your regular responsibilities. Collaborate with others and capitalize on personal relationships you have built. Seek assistance and involve your peers in the decision-making process. By making them feel invested in your success, you can foster a sense of ownership and shared responsibility. Work with your peers and engage in a brainstorming session to generate ideas. Encourage creativity and free thinking during this process. List out the pros and cons of each solution and assess the time and effort required for implementation.

Get Feedback and Buy-In

Share your solution ideas with a small group of trusted individuals who are willing to provide feedback. Listen attentively to their suggestions and iterate on your ideas to refine and improve them further. Check with the same group if the solution addresses the problem being discussed. Their diverse perspectives can enhance the quality of your solution. It is also an opportunity to advocate for your solution by highlighting its benefits, explaining why it is important to solve the problem, and emphasizing the urgency. Start by discussing your ideas with the feedback group and gradually expand your influence by seeking buy-in from decision-makers and stakeholders.

Follow Through

Depending on the scale of the problem, it may become a major program or a side project. Make sure that you have the availability in your schedule. Regardless, continue to make progress by dedicating appropriate time. If you need support from other people, ensure that they are making the right tradeoffs based on other projects and that you have buy-in from their manager(s). As long as your key stakeholders are on board and supporting you to solve the problem, continue to tackle the problem so that you get across the finish line.

Showcase the Impact

Finally, ensure that your hard work and the resulting impact are recognized. Communicate and share the outcomes of your efforts. Discuss how you arrived at the solution, the steps taken, and the people who supported you along the way. Express gratitude to those who contributed and acknowledge their role in your success.

Example: You are observing persistent product quality issues that have been impacting customer satisfaction and retention metric. By thoroughly analyzing customer feedback and collaborating with cross-functional teams, you can identify the root cause of the problem and propose effective solutions. Through their leadership and coordination, you can implement robust quality assurance processes, resulting in improved product quality, enhanced customer experiences, and strengthened brand reputation.

By adopting a forward-thinking approach, you can stay one step ahead and mitigate risks effectively. You will be seen as a leader within the organization, which may create opportunities for advancement.

INFLUENCE WITH AUTHORITY

That's not a typo! You have often heard that TPMs have to influence without authority. However, as you grow in your career and develop leadership skills, you have to have confidence and conviction in your ideas and actions. You want to give yourself authority to influence and drive change because no one else is going to give you the authority. You do not need to have direct reports to have authority. If you have been given the charter to deliver a program, then by virtue of that, you have the authority to make decisions, influence others, and drive outcomes. Your thought leadership skills will help you gain momentum on your ideas, but your self-authority and self-confidence in your influencing skills will get you the results you want. Your success in doing so hinges on your ability to collaborate, build relationships, and persuade stakeholders to align with your vision. As a thought leader, you have the ability to inspire others and drive change through innovation. You can identify opportunities for improvement, explore creative solutions, and challenge conventional thinking.

To influence others effectively, you need to articulate your ideas in a compelling way. Here are five strategies to help you influence others and negotiate win-win outcomes to make an impact.

- **Build relationships before influencing:** Building strong and trusting relationships is a prerequisite to influencing people and aligning teams. If you do not have a prior relationship, any requests will just come across as transactional. Non-personal communication is a deterrent to achieving desired outcomes. Connect with people at a deeper level and understand them. Help them if needed to build trust and credibility. Start relationships with no expectations or agenda. You never know when honest work relationships turn into lasting friendships.

- **Observe and listen before proposing changes:** When joining a new organization, take the time to understand the historical context and culture. Listen carefully to what is bothering people or what is important to them. Be curious and ask probing questions to understand the why.

- **Gather your data before building solutions:** Facts and data are essential tools for influencing stakeholders. Data-driven influencing can be especially useful when you haven't had the chance to build a relationship with the stakeholder yet. Objectively presented information eliminates room for judgment and defensiveness, making your proposals more likely to be accepted.

- **Focus on them before yourself:** When proposing an idea, the focus should be on the benefit it will provide to the person you're trying to influence. Empathy is a powerful tool, as it allows you to present your ideas from their perspective, making the proposed changes more appealing to them. By showcasing the potential benefits and outcomes, you can garner support and inspire others to embrace new ideas and methodologies.

- **Get feedback and incorporate it before finalizing the plan:** People are more likely to embrace change when they feel part of the design process. Therefore, soliciting feedback on your ideas and incorporating it into your plan can lead to more acceptance and support for the proposed changes.

Example: *If you need to convince a senior leader of your proposal, it is likely that you do not have a pre-existing relationship with them. On such occasions, use data to clearly call out the problem and its impact, and demonstrate how the solution will bring about a positive change. Emphasize the benefit to the senior leader and organization and ask for their thoughts. This will help make the case even if it takes a couple of iterations.*

PRO TIP: If you need to influence a senior leader with whom you may not have a prior relationship or clout, consider the option to deliver a preview of the message through another leader or senior stakeholder such as a manager that you have a strong relationship with. Having someone prep the conversation for you can be very beneficial.

Use your expertise and industry knowledge to propose innovative strategies and well-reasoned arguments supported by evidence and data. Influencing others is more of an art than a science. It requires an understanding of what motivates and excites people, which comes with experience and confidence. Your ability to influence will lead to high-impact results, creating positive change and driving the organization forward.

DEVELOP A GROWTH MINDSET

Remember Jane, the TPM on my team who would have benefitted from building EQ and empathy? After one of those contentious meetings, I had to give her feedback regarding the negative impact on the program. Jane became very defensive and did not accept that she had done anything wrong. When I asked her what she could do differently, she just stared back and said, "I don't know, you tell me!" Jane had what is called a fixed mindset - that nothing can be done to improve skills and abilities or that improvement is not needed at all.

In the journey of growing as a leader, one of the most powerful mindsets to cultivate is a growth mindset sometimes also referred to as open mindset. This is the belief that abilities and intelligence can be developed through dedication, effort, and a willingness to learn. It begins with a deep curiosity and a genuine love of learning. Leaders with a growth mindset actively seek out new knowledge, skills, feedback, and experiences. They approach challenges with an open mind, seeing them as opportunities for growth rather than obstacles. They are seldom dismissive of feedback and instead show curiosity in understanding which actions led to what impact.

When you believe you are always learning, no matter your expertise, you are more likely to embrace challenges and persevere through difficulties. By adopting a growth mindset, you embrace your curiosity - asking questions, seeking diverse perspectives, and exploring new ideas. You will also inspire and motivate your team members to adopt the same mindset, creating a culture of continuous learning and improvement. With a growth mindset, you are better equipped to make informed decisions, and encourage creativity and innovation within your teams.

Many companies emphasize the need to hire people with a growth mindset because those people will learn any new skill with ease as opposed to someone with a fixed mindset. These companies often use behavioral interviews to assess candidates' willingness to learn and be open to feedback. Developing a growth mindset is a lifelong journey. It requires self-awareness, intentional effort, and commitment. It also requires being comfortable with the discomfort of not knowing everything. Being closed minded or having a rigid approach can hinder your success. When faced with feedback or criticism, it is natural to feel defensive and protective of our egos. However, it is essential to recognize that defensiveness limits growth and stifles learning opportunities. Those who ask questions to learn more about the situation, impact, and which alternate actions would have been more helpful, are more likely to improve later even if they may not initially feel comfortable receiving such feedback.

Here's what you can do when receiving feedback that is difficult to hear.

- **Acknowledge and reframe:** Recognize defensiveness as a natural response but choose to reframe it as an opportunity for growth. See feedback as valuable information that can help you improve and develop.

- **Listen first:** When receiving feedback, actively listen without interruption or justification. Seek to understand the perspective of the other person and reflect on their insights.

- **Ask to learn:** Adopt a learning orientation by focusing on what you can gain from the feedback rather than defending your ego. Ask questions about the situation and talk through what actions can help.

- **Seek feedback proactively:** Actively seek feedback from trusted mentors, colleagues, and team members. Ask what you can do better or differently to make a bigger impact. Focusing the feedback on future potential can be helpful for your peers as well.

- **Reflect and iterate:** Regularly reflect on your own performance and actions. Be open to self-evaluation and recognize opportunities for improvement. Take deliberate steps to implement changes and learn from past experiences.

> **PRO TIP:** Besides your manager, ask for specific feedback from your peers and key stakeholders regularly. Don't wait for the performance review cycle, instead get feedback in real time.

Example, if you delivered a key presentation, ask 1-2 people from the audience about what worked and what you could do differently or better in delivering the presentation. This request is very specific and focuses on the future, which makes it easier for the other person to give you feedback.

MANAGE UP EFFECTIVELY

When I first attended the "Managing Up" training, which was part of the mandatory onboarding curriculum, I was intrigued. I had not heard of the term prior to that training and I didn't think it was my responsibility to "manage" my manager. However, it was really helpful to understand the value of managing up and what is required to do so.

What is Managing Up?

Managing up refers to the process of developing a positive and productive relationship with your manager and even their manager. It involves taking control of your conversations and leveraging opportunities to steer them in a direction that aligns with your long-term career goals.

Why is Managing Up Important?

Managing up is an essential part of career success as it helps to effectively communicate your career goals and aspirations. It is important because you own your career, and no one else is invested in your success more than you. While managers will invest in your success, they cannot read your mind and know your goals unless you explicitly communicate them. Therefore, it is up to you to tell your managers what you want and where you want to go. Managers are often juggling multiple priorities and working with several stakeholders. When you take charge, it becomes easier for the managers to work with you.

Managing up has a direct impact on:

- **Performance reviews:** Your manager needs to know what you have accomplished to give you a performance rating that aligns with the work you have done. By making your manager aware of your accomplishments, aligning with them on priorities, and taking ownership, your manager can give you a fair and accurate performance review.

- **Promotions:** Your manager is the best person to help you get a promotion. By effectively communicating your aspirations, you can help your manager identify opportunities that will help you reach the next level in your career.

- **Opportunities:** To get good performance reviews or a promotion, you need the right opportunities that stretch you and help you develop the skills you need for the next level. Your manager is the best person to provide you with these opportunities.

- **Building a positive relationship:** Managing up helps build trust and credibility, leading to a productive relationship with your manager. Most people say that their manager is the most important part of their work satisfaction.

How to Manage Up Effectively?

You can take concrete steps to "manage up" in a way that feels seamless and aligns with your values. Managing up doesn't mean you have to suck up and become a "Yes" person. In fact, it actually helps you create more trust in the relationship. Managing up also requires you to be comfortable advocating for yourself. It requires you to have empathy, so you can understand your manager and can effectively work with them. When you put yourself in your manager's shoes, you are likely to develop a deeper understanding of their motivations and triggers, which will help to navigate any tricky situations.

Schedule regular one-on-one meetings: Weekly cadence is preferred, but biweekly at a minimum is required, so both you and your manager feel connected to each other and can build a strong foundation. You should be responsible for driving the agenda for most of the meeting. Every quarter, you can make it a career conversation where you talk about your career goals and how you want to operate in the next quarter or two. Brainstorming ideas and discussing problems should be the focus of 1:1s. They are not meant to provide status updates.

Ask questions and connect the dots: Ask your manager lots of specific but open-ended questions. This helps you build awareness of what's going on around you, and it also helps you uncover problems. Asking questions will indicate to your manager that you are curious to learn. Share with them any information based on your knowledge from across the organization.

Solve your manager's problems: When you uncover interesting problems, ask your manager how you can help. Better, take ownership of the problem to free your manager to focus on other things. When you solve problems for your manager, it builds trust, credibility, and establishes you as a leader for them.

> **PRO TIP:** Share a 1:1 agenda document with your manager and use it to note down key decisions and discussion points. Set aside 10-15 minutes prior to your 1:1 to go over what you want to achieve in that 1:1. This intentional habit will make your 1:1s more fruitful.

Managing up is also a great way to develop a strong working relationship with your manager, build trust and credibility, and advance your career.

BUILD EXECUTIVE PRESENCE

Executive presence is a quality that sets leaders apart and enables them to inspire, motivate, and influence others. Think about a leader or speaker who you admire and enjoy listening to - Barack Obama, the former president of the United States, is known for his charismatic leadership and sense of humor. In the tech world, Jeff Bezos is recognized for his long-term thinking and bold decision-making. They have a certain presence that captivates their audience and commands attention. This presence is what you may strive to develop as you grow your leadership. It's about being able to confidently assert yourself, communicate effectively, and make a lasting impact on others.

For many, developing executive presence doesn't come naturally. It requires practice, self-reflection, and a willingness to step outside of your comfort zone. I can relate to this struggle personally. I used to be the quiet person who hesitated to speak up in big meetings. But I learned that building executive presence is within reach for anyone, including yourself. It begins with building the courage to assert yourself, which gradually transforms into confidence. It requires intentional effort and practice.

Building executive presence comes by developing and practicing many of the skills and behaviors already discussed in earlier chapters of this book. The more you practice, the more you will be confident and that confidence will amplify your presence automatically.

Become Comfortable with Discomfort

There are many situations that may make you feel uncomfortable, such as public speaking which can be intimidating for many but is a valuable skill to develop. For you, it may be speaking up in a meeting, questioning a senior leader, or challenging the status quo. If you are uncomfortable, then challenge yourself to get over the discomfort. Look for opportunities to speak up, ask questions or present ideas, whether it's in team meetings, conferences, or workshops. Start small and gradually increase the complexity and scale of those actions. Over time, you will become more comfortable and confident in expressing yourself in front of others.

PRO TIP: Take improv classes to enhance your communication skills - listening, reacting, presenting, etc. It will help you develop quick-thinking skills and become more comfortable with being in the moment.

Remember, building executive presence is an ongoing journey. It requires dedication, self-awareness, and continuous improvement. Surround yourself with a supportive network of peers who can provide guidance and encouragement.

GIVE BACK AND EMPOWER OTHERS TO GROW

I started a TPM Mentorship Circle at one company I worked at as a way to connect, share, and learn with other TPMs. This circle got so much interest from aspiring TPMs that we went on to create a focused group for Aspiring TPMs. On another occasion, I, along with two fellow TPM Managers, started a TPM Mentorship Program for the TPMs in our engineering division. After running these successfully for two quarters, the TPM mentorship structure was adopted by the entire division, which included engineering, product, and other functions.

One of the most rewarding aspects of growing as a leader is the opportunity to give back and empower others to grow. This involves actively supporting and mentoring your team members, stakeholders, and even the broader TPM community. By sharing your knowledge, experiences, and insights, you can inspire others to reach their full potential, create a positive impact, and leave a legacy behind. The best part is that you don't need to be a people manager to empower others.

Support and Mentor Team Members

You are uniquely positioned to support and mentor your team members. Create an environment such as a mentorship circle that fosters growth and provides opportunities for them to develop their skills and knowledge. Actively engage with team members by providing constructive feedback, offering guidance, and sharing best practices. Help them identify their strengths and areas for improvement, and encourage them to set meaningful goals. By investing in their growth and development, you empower them to become future leaders and contribute to the success of the organization. Actively participate in professional communities, industry events, and online forums to connect with fellow TPMs and provide guidance and support. By mentoring stakeholders and engaging with the community, you contribute to the growth and advancement of the TPM profession as a whole.

Teach Others to Fish

Empowering others to grow goes beyond providing guidance and support—it involves helping them become self-sufficient and independent learners. Instead of simply giving them answers or solving their problems, teach them how. Encourage a mindset of continuous learning and problem-solving by asking thought-provoking questions, guiding them through the process, and providing resources and tools for self-directed learning. By fostering their ability to think critically and find solutions independently, you equip them with valuable skills that extend beyond their current roles.

Create a Culture of Collaboration and Knowledge Sharing

As a TPM Leader, you can create a culture of collaboration and knowledge sharing within your organization. Foster an environment where team members feel comfortable sharing their expertise, insights, and lessons learned. Encourage cross-functional collaboration, organize knowledge-sharing sessions, and implement platforms or systems for capturing and sharing best practices. By facilitating the exchange of knowledge and experiences, you create a collective learning environment that benefits the entire organization.

Recognize Achievement and Celebrate Growth

Lastly, as you give back and empower others to grow, remember to recognize and celebrate their growth and achievements. Acknowledge the progress and milestones of team members, stakeholders, and mentees. Send out a progress report or give a shoutout in team meetings or company all-hands to celebrate successes and share the stories of growth and development. By publicly acknowledging their efforts, you create a positive and supportive culture that motivates and inspires others to embrace their own growth journeys.

By building community, you create a legacy of leadership that extends beyond your individual contributions. You contribute to the growth and success of individuals, teams, and the wider community. Through your guidance and support, you enable others to reach new heights, develop their skills, and make a lasting impact.

Embrace the opportunity to give back and watch as the seeds of growth you sow flourish in the hands of those you empower. Gil Segev did just that when he was navigating career transitions and looking to connect with other fellow TPMs. He started a small server with the intention to connect and share learnings with a few like-minded TPMs. Now that server - the TPMs Unite Discord group - has become one of the biggest TPM communities. Read more about Gil's journey at the end of this part of the book.

EDUCATE OTHERS ABOUT THE TPM ROLE

As a TPM leader, part of your growth and development involves educating others about the role and how they can effectively leverage their team to drive success. This is an ongoing process that requires empathy, patience, and a clear understanding of the value that technical program managers bring to the table. Not everyone may be familiar with this role or fully understand its responsibilities. Take the time to listen to others' perspectives, address their concerns, and provide clarity regarding the purpose and impact of the role. Here are four suggestions to get you started:

- Share how TPMs drive the successful execution of complex programs, facilitate communication between teams, and ensure alignment with the overall business objectives.
- Reference earlier chapters in this book or any relevant resources that provide a comprehensive overview of the role and help them gain a deeper understanding.

• Highlight the best practices for leveraging TPMs effectively. Explain how teams can work in partnership to maximize their impact. Emphasize the importance of clear communication and setting expectations. Encourage others to proactively engage with your team, seek their guidance, and involve them in decision-making processes.

• Share real-life examples and success stories of how TPMs have positively influenced programs, teams, and organizations as a whole. Highlight specific instances where your team has effectively resolved complex issues or driven innovation.

* * *

As you continue evangelizing the role, encourage feedback from others to gauge their understanding and adapt your approach accordingly. By actively educating others about the role, you contribute to building a culture that values and leverages their expertise, giving you a seat at the table and building representation for the TPM function.

PRO TIP: Keep a document handy that explains the role, mission, and responsibilities. You will always have new team members who may have never worked with a TPM before. Share this document with them. Be patient and open to answering questions, addressing concerns, and providing further clarification as needed.

CHAPTER 22

FINDING YOUR SUPPORT SYSTEM

If you want to go fast, go alone.
If you want to go far, go together.

– African Proverb

In the first 7-10 years of my career, I didn't realize that I needed mentors or coaches. The first time I got a coach was by chance as part of a career conversation training I took. Even though I had just a few sessions, it proved to be highly valuable in making some important career decisions. At that same time, one of my colleagues who had just joined at the time asked me about mentors and explained how she has a set of mentors outside the team and company. I could see how her intentionality in creating a support group was enabling her career. A few months later, I learned about how sponsors can enable me and how they are different from mentors. I enrolled in a sponsorship program, which proved its worth over a short span of time. Today, I can say that I am fortunate to have had good mentors, coaches, and sponsors in my career. In fact, I wish I had engaged them earlier in my career as each of them has been highly valuable in my development. I cannot emphasize enough on the importance of having a coach, a mentor and a sponsor. Think of them as your personal board of directors/ advisors. You don't just need one of each, you can have multiple mentors or coaches depending on your goals.

Before we get into why you need them and how you get them, let me describe each and the difference between a mentor, a coach, and a sponsor.

MENTOR

A mentor is an experienced and trusted advisor who imparts wisdom, knowledge, and guidance in specific areas of interest. Today, many companies recognize the importance of mentorship and have implemented mentorship programs to facilitate these connections. However, you can also take the initiative to approach individuals you admire and request their mentorship. Leverage internal resources and external networks to identify a mentor that aligns with your goals and needs. The mentor-mentee relationship begins by setting up an initial meeting where expectations, goals, and topics of interest are discussed. Once you are aligned, be proactive by preparing in advance and listening actively. As the mentee, you can choose to have the mentorship for short-term or long-term, depending on your aspirations.

PRO TIP: Engage with multiple mentors from different backgrounds, as it allows for diverse perspectives and insights.

Remember, mentorship is a reciprocal relationship based on trust, respect, and mutual learning. It is a valuable opportunity to tap into the expertise of seasoned professionals and gain invaluable insights and support. By finding the right mentors and actively engaging with them, you can accelerate your professional growth, expand your network, and achieve your goals more effectively.

COACH

Coaching, as defined by the International Coaching Federation ("ICF, the Gold Standard in Coaching | Read About ICF.", n.d.), is a transformative approach aimed at enhancing your personal capabilities, interpersonal skills, and capacity to understand and empathize with others. It is a dynamic and thought-provoking process that ignites inspiration and empowers you to unlock your full potential, both personally and professionally. Through coaching, you can tap into previously untapped sources of imagination, productivity, and leadership, enabling you to make informed choices and overcome challenges.

The coaching process is flexible and adaptable to your specific needs. It can take the form of one-on-one sessions or group settings, allowing you to benefit from individualized attention or the collective wisdom of a diverse group. Moreover, you have the freedom to engage multiple coaches who specialize in different areas, ensuring comprehensive support for various aspects of your life. Unlike mentorship, which primarily focuses on specific development goals, coaching centers around your holistic growth as an individual. It focuses on your unique strengths, values, motivations, and aspirations, enabling you to overcome obstacles and reach new heights through your own ideas and actions.

Coaching is typically a longer-term commitment, as it takes time to unravel deep-rooted patterns and achieve sustainable growth. In fact, having lifelong coaches can contribute to higher levels of personal and professional fulfillment. There exists a wide range of coaching specialties, such as leadership coaching, executive coaching, life coaching, and time management coaching, among others. Each specialization brings a distinct area of expertise that you can leverage to address specific challenges and explore new opportunities. By engaging in coaching, you embark on a journey of self-discovery, self-improvement, and self-empowerment. It is a collaborative partnership between you and your coach, built on trust, confidentiality, and mutual respect. Through insightful conversations, introspective exercises, and practical strategies, coaching equips you with the tools and mindset needed to enhance your leadership skills, take proactive steps towards your goals, and lead a more fulfilling life.

Coaching is a powerful developmental approach that nurtures personal growth, enhances interpersonal skills, and facilitates self-awareness. It embraces your unique strengths, values, and motivations, empowering you to overcome challenges, unlock your potential, and achieve both personal and professional success. With the support of a skilled coach, you can embark on a transformative journey of self-discovery and continuous growth, leading to a more fulfilling and purposeful life.

Identifying coaches who are the right fit for you is a crucial step in your personal and professional development. You can seek recommendations from friends or colleagues. Similar to mentorship, you have to take ownership of your actions, communicate openly, and provide feedback to the coach. This will help you get the most out of your coaching engagement.

Remember, coaching is a collaborative partnership that empowers you to reach your full potential. By selecting the right coaches and actively engaging with them, you can gain valuable insights, develop new skills, and achieve your personal and professional aspirations.

SPONSOR

While mentors and coaches play crucial roles in personal and professional development, sponsors hold a distinct position when it comes to advancing your career. Sponsors are influential individuals within an organization who can go beyond offering advice and actively advocate for your advancement and success. A sponsor is more than just a mentor or a well-wisher. They are senior leaders or influential individuals who have the power and influence to help shape your career. Sponsors can provide you with opportunities, visibility, and access to critical networks that can propel your professional growth. They act as your champions and vouch for your abilities, advocating for your promotions, assignments, and recognition.

Unlike mentors, who typically provide guidance and advice, sponsors take a more active role in your career advancement. They are the ones who can open doors, recommend you for challenging projects, and provide public endorsements of your skills and potential. They are instrumental in helping you navigate the complexities of the organizational landscape and positioning you for success.

Seeking sponsors is a strategic move for ambitious professionals who are committed to accelerating their careers. By cultivating relationships with sponsors, you gain access to their expertise, influence, and valuable insights into the inner workings of the organization. They become your allies, helping you overcome barriers, navigate political dynamics, and seize growth opportunities.

Example: You are a mid-level engineer who's highly technical and comes up with great architectural solutions. However, you are a bit shy or introverted and find it difficult to talk about your achievements or speak up without being prompted. You want to get promoted to the next level. While you can leverage a coach to build your leadership skills, a sponsor can help be your voice by highlighting your work, finding your opportunities to present, or making space to help you speak up in meetings. Your manager may not always be in every meeting, so having another person be your voice can be very helpful.

Before identifying sponsors, make sure you are demonstrating excellence in your work. Actively seek opportunities that provide you broader visibility throughout the organization and position yourself as a subject matter expert. Find opportunities to build authentic and mutually beneficial relationships with stakeholders from across your organizations. Once you find a sponsor, work with

244 | THE *ART* OF STRATEGIC EXECUTION

them closely to understand how you can continue to grow, and share your accomplishments and aspirations.

Understanding the unique role of sponsors and actively seeking their support can significantly enhance your professional journey, leading to increased visibility, advancement, and opportunities for success.

DO YOU NEED ALL THREE?

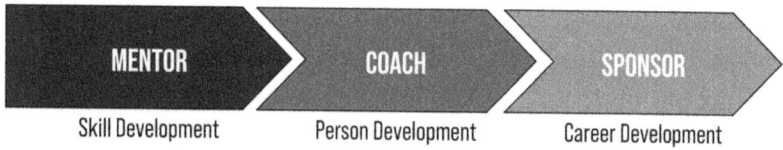

MENTOR	COACH	SPONSOR
Skill Development	Person Development	Career Development

The Difference between a Coach, Mentor, and Sponsor (Adapted From (Grove 2019))

A mentor talks with you, a coach talks to you,
and a sponsor talks about you.

(Osmani 2022)

Your life is multifaceted and you need different perspectives for each of them. You may be strong in some areas while needing more support in others. Having a mentor, coach, and sponsor in your professional journey can provide a comprehensive support system that maximizes your growth and potential. Each role brings unique benefits that complement and enhance one another. Together, these three roles create a well-rounded support network that combines mentorship, skill development, and influential backing, enabling you to thrive and succeed in your professional endeavors.

Example: *You are a professional - say a TPM :). You have just joined a new company - your dream company and job. You were previously at a small company on the opposite coast and recently moved to take this job. You are feeling a bit overwhelmed by all the changes, but you want to succeed from the get-go in your new job. What do you do?*

With so many things going on, this would be a great time to build your board of advisors consisting of mentors, a sponsor and a coach. You can look for mentors in your new team or company. These mentors can help you with adjusting to the new company culture and help you build technical skill sets. You can hire a professional coach either within or outside your company to help you deal with change management and address the feeling of being overwhelmed. The same coach or a different one can also help build your confidence and leadership skills that will enable you to navigate your new role with ease and influence executives. And you can start looking for a sponsor that will help with your career goals. You don't need a sponsor from day one at the job, but you can be on the lookout. Eventually, this sponsor can support your promotion in a calibration review meeting with your manager.

* * *

It doesn't matter whether you are young or old, new graduate or experienced professional, single or married, there will always be situations and challenges where having one of these people in your network will prove invaluable. Why struggle through it alone when you can get help? These partnerships don't always have to cost money and even if they do, think of it as an investment into your life and career, the benefits of which you will continue to reap lifelong.

PRO TIP: Interview several mentors, coaches, and sponsors to learn about their style, clarify expectations and understand what a successful relationship may look like for you.

📣 TPM SPOTLIGHT

GIL SEGEV, SENIOR STAFF TPM

Gil has been working as a TPM for the majority of his career, spanning nearly 20 years. He began his journey at a big semiconductor chip manufacturer as a developer, focusing on databases for IT. He always had a passion for technology and the challenges it presents and realized early on that the influence and customer interaction were closely tied to project management. So he started venturing into project management roles and over time gained a wealth of experience. As projects grew into larger programs, he transitioned into program management. He also had the opportunity to work on various teams, including firmware and software development, and later, hardware, where he managed deliveries for numerous PC launches. Eventually he moved to working at a big internet company.

Can you elaborate on the differences you've observed between the TPM role at Intel, Amazon, and Meta?

Gil: The differences are like night and day, and it all comes down to company culture, which has a significant impact on the TPM role. In mature companies where the culture is more top-down, TPMs are often given a clear charter, substantial decision-making power, and influence. In contrast, newer companies, with a bottom-up engineering culture require TPMs to navigate their own path. There's often no clearly defined charter, so building personal relationships, establishing influence, and demonstrating technical acumen become essential.

Is one of these environments more challenging than the other?

Gil: It really depends on your strengths. The newer, bottom-up culture may seem less structured and more ambiguous, as engineering often defines everything, including scope. In top-down cultures, there's typically more ownership for TPMs. So, while there may be challenges in both, the nature of those challenges differs.

What did you need to adapt when transitioning between these different company cultures?

Gil: I had substantial credentials having worked as a TPM for many years, but I had to shed all preconceptions about what a TPM does. I had to admit that I didn't fully understand the culture at each new company. So, I walked in with an open mind, willing to start from scratch in building influence and relationships. Communication and decision-making were particularly challenging in these transitions. I also had to adjust my approach to organization and methodical thinking. My first company was highly structured, while my current company embraced more fluidity, which meant building a reputation where the role lacks a clear charter, was more challenging. In essence, I had to take ownership of defining my role and scope, which was both challenging and rewarding. Promotions are also very different at these companies. There is often a misconception that doing the job well for a time period can get you promoted but this is not true for some modern tech companies. You have to over-deliver on scope and go above and beyond and take risks.

What surprised you the most as you navigated the TPM role?

Gil: The most surprising aspect was the sheer amount of knowledge I had to acquire, and it wasn't just technical knowledge. It was about understanding a broad range of information and having the ability to control and make decisions effectively. At large companies, with thousands of employees and diverse programs, being able to grasp the big picture while diving deep into details was crucial. I learned that it's impossible to know everything, but knowing how to find the right information and make informed decisions is a skill that proved invaluable.

What do you like most about being a TPM?

Gil: The biggest challenges are the most rewarding aspects of being a TPM. I enjoy having a panoramic view of projects and programs, something even engineers often don't have. As a TPM, I get to work with multiple moving pieces, see the entire map of where we're going, and manage the journey to the end state. It broadens my perspective and keeps me engaged.

On the flip side, what do you dislike about the TPM role?

Gil: I see documentation, processes, and notes as necessary components of the role, but they can easily be misunderstood or abused. Some people perceive TPMs as primarily responsible for creating processes and taking meeting minutes, which misses the point. Processes should be viewed as a means to an end, not the end itself. TPMs are here to deliver actual value, not just manage tasks.

What do you think is the best way to leverage TPMs in an organization?

Gil: The best approach depends on the company's culture. In some cases, having TPMs embedded within engineering teams, where they can be hands-on and deeply involved, is most effective. However, balance is key; if TPMs become too detached from engineering, they might end up just managing tasks and processes.

Can you share your experiences transitioning from an IC6 to an IC7 role?

Gil: The transition from IC6 to IC7 is more about broadening your scope and taking initiative. You have to reach out and do more, aligning your work with IC7-level responsibilities. It's about defining your archetype as a TPM, whether you're a technical specialist, a generalist, or something else. IC7 requires you to understand the vision of directors and VPs, provide them value, and actively solve their top challenges. It's about building relationships, expanding your scope, and being willing to take on tasks others might shy away from.

You've stayed on the IC path rather than pursuing a management path. Why?

Gil: I did consider the management path multiple times, and there's no right or wrong choice here. It ultimately depends on what you enjoy doing. Whether it's solving engineering problems, working on infrastructure, or enabling other people and building their skills, both paths offer opportunities for impact. The IC path has the advantage of limitless scope, even though there may not be many IC9 roles. Compensation and other factors are often similar on both tracks, so it's a matter of personal preference.

Tell us about your motivation behind starting the TPM community on Discord.

Gil: I launched the TPM community on Discord in 2021, which was during the height of the COVID-19 pandemic. I found there were plenty of online resources for various other functions, but not enough for TPMs, especially at senior levels. Blind was one of the few platforms where TPMs could connect but it wasn't interactive. So, I started a small Discord server with the idea of bringing TPMs together to discuss and share their job seeking experiences. Over time, it grew beyond just interviewing; now it's a platform for discussing skills, mentoring, job opportunities, and more. The timing was right, and the community has been incredibly rewarding, fostering connections and helping others.

What is your hope for the TPM community's future?

Gil: My hope is for the TPM community to become a shining beacon for everything related to TPMs. I'd like it to be a central hub for skills development, mentorship, job searches, and recruiting. I want TPMs to take a more prominent role, as they often operate in the background. Together, we can learn from each other and help each other succeed.

Finally, what advice would you give to other TPMs or aspiring TPMs?

Gil: Based on my experience, I'd say that those who actively seek change and embrace challenges tend to advance more. Not everyone may want to change or advance, but if you're looking for growth, don't hesitate to make shifts in your career—change jobs, companies, or even industries. Meteoric rises often occur when you're willing to make those changes. It's not just about titles and compensation; it's about gaining diverse experiences and continuous learning.

PUT IT INTO PRACTICE

✓ Meet with at least one new person every week. If you are in a new team, ask every person to name three other people to meet and go meet them.

✓ Rate your leadership skills and identify which areas you need to grow in most based on your long-term career goals.

✓ Identify situations which make you most uncomfortable. Now identify, what ideally would you do if you were not afraid? Write down how you want to get over the discomfort to build specific skills.

✓ Role play with a friend or trusted colleague to practice situations where you need to build skills. Example: If you need to influence a senior leader to get buy-in on your proposal, ask your friend to act as the leader and you act as the TPM. Think like the leader as you practice what you would present.

✓ Create a document for yourself to identify your personal board of directors. Assign each of them roles - mentor, advisor, coach, sponsor, etc. and tag each with the area or skill they can help you grow in. This will help you see where you may need additional support.

Part 5

GETTING INTO TPM
PEOPLE MANAGEMENT

*Management is doing things right,
leadership is doing the right things!*

– Peter Drucker

CHAPTER 23

IS PEOPLE MANAGEMENT RIGHT FOR YOU?

The first rule of management is delegation.
Don't try and do everything yourself because you can't.

– Anthea Turner

Throughout my TPM career, I often found myself in the role of the first or only TPM, giving me the opportunity to collaborate closely with senior leaders on the overall organizational strategy. This involved not only growing the TPM team and defining its responsibilities but also advocating for the TPM role. When I began mentoring other TPMs, I discovered a passion for helping in their professional growth and I continued creating opportunities for TPMs to connect with each other. I thoroughly enjoyed these aspects of the work, and it seemed like a natural progression to transition into people management. Growing into the people manager and eventually an organizational leader, I understood that people management goes beyond managing people. There is much more complexity, which requires one to approach the role differently. I still enjoy working with people and helping them grow, which is one of the reasons I became a coach and also the reason you are reading this book.

Today, the TPM role has evolved, where management may not be the only path. There are a multitude of opportunities within the IC track and other intriguing career paths adjacent to TPM, beyond just product or engineering. However, there is still the misconception among TPMs that people management is the sole means to progress in their careers. When individuals approach me seeking guidance on transitioning into a people manager role, my initial inquiry is, "What aspects of people management resonate with you?" Often, the responses revolve around the desire to delegate tasks and scale their impact. Some share their motivation to fix issues they couldn't address as an IC, or to be the kind of supportive manager they lacked in the past. While these aspirations are valid, they alone may not suffice for success in a people management role. The issues you face as a manager will be different in nature and you still may not be able to "fix" all the issues. Not everyone naturally gravitates towards people management or may enjoy being a people manager, and that's perfectly acceptable.

Stepping into people management is a significant career milestone and the journey from being an IC to a manager of TPMs brings about both excitement and challenges. However, managing a TPM team can be uniquely challenging. Unlike engineering managers, TPM managers are likely to have a bigger scope in terms of breadth across teams since the ICs on the TPM are working cross-functionally. For instance, in my first manager role, my team of six TPMs was working across five different products and platforms and partnered closely with a 200-person engineering organization in addition to partner team stakeholders. TPM managers have the extra responsibility of advocating for the role since it is not always well understood like engineering. Regrettably, I've observed numerous TPM leaders who struggle to champion their teams or cultivate trust when collaborating with stakeholders.

While expertise in technical and program execution remains incredibly valuable, this transition necessitates a fundamental change in mindset. The focus of a people manager shifts from personal achievements to fostering the growth and development of others. It also demands a profound understanding of human dynamics and effective leadership at all levels. Recognizing and embracing these essential shifts enables individuals to make informed choices aligned with their passions, strengths, and long-term professional aspirations.

MINDSET SHIFT: TRANSITIONING FROM IC TO MANAGER

Your Impact is Dependent Upon Others

As a TPM manager overseeing a team of TPMs responsible for programs across the organization, it can feel disempowering to relinquish control and trust your team to drive successful program outcomes. You might be accustomed to directly managing programs as an IC and may struggle to delegate decision-making and ownership to your team members. As a people manager, your impact is no longer solely determined by your individual contributions and capabilities. Accepting that not every outcome is within your control can be difficult at first, but know that it is required for long-term success. Becoming a people manager requires shifting the focus from individual achievements to collective success. If you find yourself hesitant to trust others or struggling to let go of control, it may indicate a need for growth in this area. Take some time to reflect and talk to other managers to learn more about how they navigated such transitions. The key is to have patience with yourself and others through this learning period. Leaning on your IC leadership skills discussed previously like building accountability with others, delegation, and setting clear expectations will come in handy.

People First Approach

Balancing the needs and development of your TPM team members while driving multiple programs yourself can be challenging. You may struggle to prioritize their growth and well-being amidst competing program deadlines, stakeholder demands, and your own career aspirations. Developing a people first approach requires empathy, active listening, and a genuine interest in supporting the growth and success of others. It will help you cultivate high-performing teams that deliver results.

As a manager, you also need to understand that your team may be dealing with a range of problems, personal or professional, that can impact their performance and interpersonal behavior. These conversations can be tough and uncomfortable, and there is a fear of conflict. It's understandable that addressing sensitive topics is difficult as stakes are often high for both parties. The ability to navigate crucial conversations is an essential skill for people managers and it requires assertiveness, empathy, and effective communication skills. Showing care and providing support in tricky situations can help your team members navigate the circumstances and continue to deliver on their goals.

MANAGEMENT RESPONSIBILITIES

Being a people manager involves not only leading and inspiring a team, but also handling various administrative responsibilities related to performance management, compensation, human resources, employee relations, and employee engagement. These administrative duties can often take up a significant portion of your bandwidth, so you have to be mindful of how you manage your time. Managing these administrative yet necessary aspects helps create a positive work environment and ensures the success of both the team and the organization. While most of these responsibilities are common to any people manager role, there are some nuances that are important for TPM managers to understand due to the nature of the TPM function and how it is seated within an organization.

Performance Management Cycle

The performance management cycle typically includes goal setting, ongoing feedback and coaching, mid-year reviews, and end-of-year performance evaluations. Many organizations also go through the lengthy process of calibrations, which helps the manager and their counterparts provide a fair and unbiased performance review.

For TPMs, much of the feedback comes from partner functions like product and engineering, so it's especially important for TPM managers to maintain open lines of communication with other leaders to understand how your team members are performing. You have to align your team and cross-functional partners with a clear evaluation criterion to ensure fairness in the process. There can be additional challenges due to potential biases against the role and varying program contexts. Collecting feedback for all your direct reports, collating it to create a cohesive message, and delivering the feedback itself will take time and effort. Sometimes, performance reviews have to be completed within an aggressive timeline while you are already juggling other critical programs.

Compensation Philosophy

Managing compensation involves overseeing the process of determining salaries, bonuses, incentives, and other financial rewards for team members. It requires staying informed about compensation trends, market rates, and internal policies to ensure fair and competitive compensation packages. You will need to work with your recruiting and HR partners to ensure a fair compensation process.

HR and Employee Relations

Handling various personnel-related matters, such as onboarding new hires, managing leaves of absence, addressing performance issues, resolving conflicts and even managing out low performers, requires a deep understanding of HR policies, procedures, and legal considerations.

Employee Engagement

Many mature companies conduct a twice-a-year employee engagement survey to understand how the employees are feeling about the company, their manager, and their work. These surveys are mostly anonymous, so that you are more likely to receive open and honest feedback. Once the results are out, you are responsible for solving any systemic issues based on the feedback received. You may be asked by management or your stakeholders to own certain initiatives or you can also take on action items that will address the feedback from your team. Improving certain scores becomes a goal for many managers, which means you will be responsible for helping improve your team scores.

* * *

Becoming a manager is a weighty responsibility and can sometimes be more work than an IC at the same level. At the same time, it can also be incredibly rewarding and impactful.

PRO TIP: Check the book "Crucial Conversations: Tools for Talking When Stakes are High" to help you navigate tough conversations with more grace.

CHAPTER 24

BECOMING A TPM PEOPLE MANAGER

*Keep your fears to yourself, but share your courage
with others.*

– Robert Louis Stevenson

I was fortunate to have moved into management at a company that puts in a lot of effort to support managers the right way. I had the opportunity to be part of a program called "Pro-M" - short for Prospective Manager - where one receives similar training given to new managers but in a smaller cohort. This program ran for 12-16 weeks during which I had a dotted line report to "practice" management. It gave an aspiring manager like me a taste of management along with feedback without having to fully make the shift. This program helped lay the groundwork for the people management stage of my career. I started shifting my mindset and identifying what new people management related skills I needed to develop. Unlike engineering managers who may manage teams focused on one part of the stack, as a TPM Manager, I had a large scope that cut across multiple teams, which meant I needed to broaden my view even further and be aware of what was going on in areas my team was involved in. When I led the TPM organization at a startup, my team of 40+ TPMs was collaborating with an engineering organization of 1200+ engineers. That level of breadth requires a manager to be strategic while keeping a pulse on different areas of the organization.

As you think about management, start building a broader perspective of your organization and its programs beyond what you are already managing. Here are a few strategies that will help you successfully transition to people management and make the right first impression as a people manager.

THINK STRATEGIC, THINK BIG

In the months leading up to my transition to people management, I had frequent conversations with my manager about the programs and organizational needs as we had just been reorganized into a separate division. One thing I heard over and over again was that the head of our organization was always concerned about understanding if the work being done by the teams was aligned to the top level goals. I took this opportunity to dive deep and learn more about why the leadership team needed this visibility and what problems it would help solve. The strategic approach to solving this problem helped me uncover other needs like aligning our yearly plans to other organizations and reporting further up the company to the CEO. I had an idea that a new planning framework implemented from the ground up would solve the problem, so I started working cross-functionally with my engineering and product stakeholders and got their feedback. I also connected with other teams who had implemented a similar framework. Besides, coming up with the solution, I drove the implementation end to end with the entire organization of 300+, creating training material and customizing internal tools. I iterated taking proactive ownership and solving a strategic organizational problem made the case for my move to people management stronger.

Think beyond your programs and turn your focus to the organization's effectiveness. Solve gnarly problems that affect the overall program delivery and collaboration dynamics of the entire organization. As a senior member of the team, you possess valuable insights into the workings of programs and can leverage this knowledge to address larger organizational challenges.

- **Understand the broader context:** Gain a deep understanding of the organization's goals, strategies, and challenges beyond your program. This broader perspective will enable you to identify areas for improvement and contribute to the organization's overall success.

- **Analyze root causes:** Develop the ability to identify underlying causes of organizational problems and make data-driven decisions. Use your analytical skills to assess risks, anticipate obstacles, and propose effective solutions that align with the organization's objectives.

- **Collaborate across teams:** Foster collaboration and establish strong relationships with other teams and stakeholders. Leverage your network to gather diverse perspectives and build coalitions to drive change.

PRO TIP: In 1:1s with your manager, ask questions that give you clues about the problems they are facing. These problems are often at the organizational level and can help you gain visibility during quarterly planning, setting up a new onboarding program, etc.

BUILD ADVOCACY

As an IC, managing up is key to advancing your career. To transition successfully into a TPM People Manager role, in addition to managing up, garnering the support of key stakeholders within your organization is essential. Your ability to cultivate meaningful relationships with key stakeholders effectively (navigate demands, address concerns, incorporate feedback, etc.), both internally and externally, will significantly impact your success. Keep an open line of communication with stakeholders to share progress, challenges, and successes. Manage conflicts swiftly, find solutions

264 | THE *ART* OF STRATEGIC EXECUTION

to competing demands and seek win-win solutions that benefit both your team and the stakeholders involved. Highlight the value your team brings to the organization and ensure alignment with broader goals. Clearly articulate how your team's efforts contribute to the organizational mission. When stakeholders see the tangible value and alignment, they are more likely to support you in your role as a TPM People Manager.

These connections could act as advocates for your career progression into a TPM People Manager role. Managing stakeholders, sometimes referred to as "managing sideways," entails skillfully navigating stakeholder demands, addressing concerns, and incorporating feedback to demonstrate your team's capabilities and your suitability to lead them. As a people manager, you become the primary point of contact for all stakeholders regarding your team's responsibilities, which means your relationships with them matter. Even your future team will benefit from your strong connections.

Here are four strategies to successfully build advocacy to achieve your goals.

- **Participate in cross-functional initiatives:** Actively participate and contribute your expertise to large multi-organizational initiatives. Demonstrate consistent and high-quality work, allowing influential individuals to witness your capabilities, dedication, and potential as a future TPM People Manager.

- **Offer support:** Collaborate with cross-functional leaders on their initiatives and show genuine interest in their work and objectives. By being a reliable and supportive team member, you can establish trust and showcase your ability to work effectively with diverse teams and leadership.

- **Participate in leadership development programs:** These programs often provide opportunities to interact with senior leaders and potential mentors from another team. Use these platforms to showcase your enthusiasm for growth and your dedication to taking on leadership roles.

- **Practice self advocacy:** Be vocal about your contributions and positive impact to the organization. Remain authentic and continue to build trust and mutual respect.

Contrary to a common misconception, a manager's success is not solely about leading their team, but also effectively working with stakeholders, other manager peers, and your manager's peers.

PRO TIP: Seek mentorship and coaching from experienced TPM managers or leaders within your organization who can provide guidance and insights.

ALIGN WITH YOUR MANAGER AND SKIP MANAGER

Securing a promotion or transitioning into a people management role often hinges on the support and alignment with your current manager and your skip level. Additionally, the availability of a people management position is contingent on the overall team size and scope. It's vital to establish a strong alignment with your management chain and cultivate a shared understanding of expectations and career objectives. Openly discussing your career aspirations and showcasing your problem-solving abilities will bolster your visibility, foster trust, and unlock opportunities for professional advancement. Scheduling regular career discussions - ideally quarterly - provides an avenue to seek feedback, make necessary adjustments, and ensure your work aligns with your objectives.

Here are actionable strategies to effectively align with your manager and skip-level for promotion to a people management role:

Transparent communication and goal alignment: Engage in open and transparent communication with your manager regarding your career goals and aspirations. Ensure these align with the expectations your manager and the organization have for the role you're aspiring to. Clarify any discrepancies and establish a clear path to achieve these goals. Regularly communicate your progress and proactively seek feedback to ensure you are meeting their expectations.

Proactive problem-solving and value addition: Proactively identify challenges and opportunities within your role and propose potential solutions. Showcase your ability to navigate complexities, offer informed recommendations, and contribute value to the team and organization. Highlight instances where your proactive problem-solving had a positive impact.

Regular check-ins and feedback: Schedule regular check-in meetings with your manager to discuss your progress, seek feedback, and align your efforts with the broader team objectives. Actively listen to their insights, address any concerns, and implement feedback constructively to enhance your performance and contributions.

Initiative and ownership: Develop the ability to anticipate your managers' needs and provide solutions or recommendations in advance. Take ownership of your projects and responsibilities. Showcase your ability to lead and guide others, even before officially stepping into a managerial role. This proactive approach highlights your readiness and willingness to embrace a people management position.

Guidance and support: Establish a strong rapport with your managers and senior leaders. Regularly engage in discussions about your career aspirations, seek their guidance and advice, and leverage their expertise to enhance your leadership skills. Find people management-related opportunities to build expertise and demonstrate readiness for the leadership role.

* * *

With these strategies, you can navigate your career with more intention. Embrace the challenges, seek opportunities for growth, and continuously refine your leadership capabilities to become a TPM People Manager. Transitioning from an IC TPM to a TPM Manager is a journey that requires continuous learning, patience, adaptability, and a willingness to embrace new responsibilities.

PRO TIP: Seek out opportunities to lead and mentor junior team members to gain hands-on leadership experience before moving into the role formally.

CHAPTER 25

MASTERING PEOPLE MANAGEMENT

Not all good leaders are good managers, and not all good managers are good leaders.

– Unknown

"People leave managers, not companies" goes the saying and you have probably left a job because of a bad manager. According to multiple research studies by Gallup (Nolan 2017), "*One in two employees have left a job to get away from a manager and improve their overall life at some point in their career, according to Gallup's State of the American Manager report. Bad managers are abundant, and they leave an impression - that's what makes movies like Office Space so funny. Almost everyone can relate to the sense of dread about coming to work when a manager makes an otherwise good job feel like a dead end.*"

This is a longstanding trend and progress is being made, but the problem still persists for various reasons. Some companies just don't have the right culture, others may have the wrong incentives, and yet others don't even provide the basic training to new managers to help them understand the impact of their responsibilities.

As with any new role, transitioning into people management means you are starting at the bottom of a new learning curve. It can get overwhelming pretty quickly as new responsibilities are

added. Your new team is expecting a lot of answers, you might be
in the middle of a big crisis, or you need to grow your team from
scratch and there are more programs than people on your team. I
hear from many new managers how they are struggling with the
increased expectations and demands for time from every part of
the organization. They think they are being good managers to their
team, but they are still not seeing the results of their impact. Most
folks who get into people management mistakenly believe that
being a people manager is just about managing the team. Being a
good manager is not enough. You have to be a good leader that can
move the organization forward and is trusted by colleagues. You
have to be able to navigate the organizational complexities beyond
the team in order to see results that matter. I strongly recommend
that you seek support from your immediate manager, mentors, and
advisors who have been on this path.

LEADERSHIP BEHAVIORS FOR MANAGERS

Leadership and management require different skills, but they
work together in a complementary fashion. Companies need both to
succeed. Good managers need to be good leaders to find long-term
success. People are generally inspired to follow leaders. If you look
at some of the most famous business leaders, most people mention
their vision, strategic thinking, or other leadership behaviors that
set them apart. No one talks about their people management skills.
Those are important to have, and it is also important to understand
the differences between leadership and management, the qualities
that define a good leader, and why you should focus on developing
leadership skills first.

Leadership is defined as the action of guiding a group of people
or an organization. It is an action-oriented role where the focus is
on what the group should be doing. Management is defined as the
process of dealing with people or controlling people. It is a process-
oriented role where the emphasis is on planning, organizing, and
coordinating resources - or how to produce desired results.

> LEADERSHIP = WHAT
> MANAGEMENT = HOW

To master people management, you must develop and exhibit various leadership behaviors in addition to manager behaviors (discussed in the next section). These behaviors not only inspire and motivate others, but also create an environment that fosters growth, collaboration, and success. These behaviors are also required as you strengthen leadership skills as an IC as discussed earlier in Chapter 21. Those leadership behaviors like taking ownership for solving complex problems, influencing, managing up, and a growth mindset become even more important. Here are six leadership behaviors that you should adopt to excel as a people manager:

Build a vision. Create a compelling vision and mission for the future and inspire others to rally behind it. Paint the future and communicate with clarity and enthusiasm. This will enable you to drive progress and ignite a sense of purpose that will motivate your team. For TPM leaders, your vision can include how your team will drive key business outcomes and/or how it will be positioned within the organization to have maximum impact.

Hold yourself accountable to your team: As a TPM leader, you have to take ownership of decisions and actions, even in the face of ambiguity or uncertainty. Hold yourself and your teams accountable for the outcomes and demonstrate a willingness to make difficult decisions when necessary. Be accountable to your team because your behavior impacts them significantly and be willing to take responsibility when something goes wrong. When you do that, it instills confidence and trust in the people around you.

Build trust: Invest time and effort in understanding your team members as individuals, showing genuine care and support. Be transparent and authentic in your interactions with your team members and be willing to be vulnerable. When you come across as an authentic leader, you create an environment of trust where team members feel safe to take risks, share ideas, and collaborate effectively.

Lead by example: Understand the importance of continuous improvement and personal growth. Seek feedback from your team members, and be willing to act on that feedback and learn from mistakes. If you want your team members to develop certain skills or behaviors, leading by example can inspire your team to embrace a growth mindset.

Embrace inclusivity: Effective leaders create an inclusive environment where everyone's opinions are valued and individuals feel heard and seen. As a TPM leader, you must actively seek diverse perspectives, encourage open dialogue, and create opportunities for all team members to contribute. This allows you to harness the collective intelligence and creativity of the team for broader impact.

Value others with empathy and humility: Great TPM leaders value each team member's contributions, strengths, and perspectives. By recognizing and appreciating the unique talents and experiences of individuals, you foster a culture of respect, collaboration, and mutual support.

PRO TIP: Check out the Situational Leadership model (The Center for Leadership Studies, founded by Dr. Paul Hersey, n.d.) to understand how you can adapt your leadership style and actions in any given situation.

Prioritizing these leadership behaviors will help you excel in your role, create a positive work environment, and drive exceptional results. These leadership behaviors serve as a foundation for effective management and contribute to personal and professional growth as a TPM Manager. As a great leader, you can inspire the next generation of leaders to do the same.

MANAGER BEHAVIORS TO CREATE HIGH OUTPUT TEAMS

Lori Goler, head of HR at Facebook said, *"At Facebook, people don't quit a boss — they quit a job. And who's responsible for what that job is like? Managers"* (Goler, Gale, and Grant 2018). So even if a manager is not "bad" per se, they are still responsible for making the job interesting enough for people to stay and empowering people to leverage their strengths and help them grow in their career. This is why being a manager is a hard job. It requires a delicate balance. At many companies, managers are evaluated on key manager behaviors focused on not just driving business impact but also ensuring that the team feels supported. These manager behaviors help cultivate high-performance teams, which is important for the long-term success of a manager.

While many of the manager behaviors outlined are not specific to the TPM function, TPM leaders generally operate in a matrixed environment, which makes these behaviors even more important for navigating tricky organizational perceptions and delivering business impact. Focusing on these behaviors also creates an environment that fosters trust and motivation within your team, leading to peak performance and broader trust with senior leadership.

Let's explore seven key manager behaviors that will enable you to build high-output teams:

Set Clear Expectations

Defining goals, objectives, and performance standards can provide clarity and direction. Articulate clearly to team members what is expected of them and by which date to set them up for success. Providing them with relevant information and guidance will enable them to meet the expectations. Brainstorm ideas with them and encourage them to make their own decisions and come to you if they are blocked. This will also help hold your team members accountable for the outcome and ensure that everyone takes ownership of their work.

Recognize Achievements

Appreciating the efforts and achievements of team members is essential for building morale, motivation, and a positive work culture. Recognize team members who meet or exceed expectations, reinforcing a culture of high performance. Understand what recognition means for each person on the team and adjust accordingly. For some folks, public recognition is motivating and for others, a financial incentive is a sign of recognition.

Match People to Their Strengths

To keep team members motivated and engaged, you should be attentive to their strengths and talents. Understanding each team member's skills, interests, and capabilities allows you to assign tasks that align with individual strengths. By leveraging their strengths, team members can contribute their best work and experience a sense of fulfillment and accomplishment. Take the time to get to know your team members and their unique strengths and consider the same when assigning programs, ensuring that team members are working on initiatives that align with their expertise. This approach not only boosts team members' motivation, but also enhances the overall productivity and quality of work.

Show Care and Provide Growth Opportunities

Managers who genuinely care about the well-being and professional growth of their team members create a supportive and nurturing environment. Take the time to listen to your team's concerns, provide guidance, and offer opportunities for learning and development. Regularly checking in with team members to understand their career aspirations and personal growth goals demonstrates care. Provide opportunities for skill development, such as attending conferences or participating in relevant training programs, and encourage team members to take on new challenges and stretch assignments to expand their capabilities and advance their careers. By investing in your team members' growth, managers foster loyalty, engagement, and a sense of belonging.

Provide Timely and Honest Feedback

It is your job as a manager to give feedback to your team members because you have the best view of each team member's performance and you are probably also hearing from their stakeholders. Providing timely and honest feedback helps individuals understand their strengths, areas for improvement, and progress towards goals. This can be done through regular one-to-one meetings or specific career conversations. Deliver feedback with objectivity, compassion, and a focus on constructive solutions. Offering specific examples and actionable suggestions for improvement while recognizing their achievements will help your team be more willing to act on the feedback. Keep in mind that feedback is a two-way street, which means you should also regularly seek feedback from your team. Creating a culture of continuous improvement will foster individual growth and contribute to overall team success.

Unblock the Team and Help Them Build Relationships

As a manager, you play a crucial role in moving your team forward. Conveying important information and helping your teams build relationships with key stakeholders can enable them to make progress faster. Actively communicate context and changes from the company to keep the team well-informed. Proactively address any roadblocks or challenges that may hinder your team's progress. You also need to keep your stakeholder informed on the work being done by your team, so they get a cohesive view of multiple critical programs across the organizations, which may not always be easily spotted by business or product teams working in specific areas.

Become Coach-Like

If you want to build high performing TPM teams, you will likely hire the smartest and the best TPMs you can get. These folks can operate independently with little guidance, but as a manager, you still need to support their career development. These folks also don't need to be told exactly what to do but they want to explore the possibilities. This is where adopting a coaching approach can evoke transformation and growth in your team members. Engaging in coach-like conversations means you don't necessarily offer solutions or answers right away, but ask questions that help your team members get to the answer by themselves. You can also encourage self-reflection, help team members identify their strengths and areas for improvement, and collaborate on action plans to enhance performance.

A coach-like manager can empower team members to take ownership of their development and maximize their potential.

> **PRO TIP:** Check out the book "The Coaching Habit" by Michael Bungay Stanier to develop coaching skills like asking powerful questions and active listening.

Cultivating manager behaviors in a structured manner can have a profound impact on motivating teams, retaining high performers, and reducing attrition, all of which will eventually result in achieving exceptional results as a people manager.

MEASURE YOUR IMPACT

As mentioned in Chapter 26, your impact as a manager is defined by the impact of your team and the impact of the organization you support. The actions you take to contribute to this impact will define your own success. When setting goals as a manager, specifically call out programs you will lead to create an impact on these entities.

Org Impact

As a TPM Manager, you can leverage your unique vantage point to make an impact on the organization beyond your TPM team. How did you align your team's expertise and capabilities to key organizational initiatives, such as launching new products, expanding into new markets, or improving operational efficiency?

What strategic initiatives did you lead that reduce inefficiencies or bottlenecks for the teams and help them collaborate more seamlessly?

Example: Implementing a new planning framework, execution best practices, etc.

Team Impact

As a TPM Manager, your primary responsibility is to foster the growth, development, and success of your team. How did you increase your team's impact or effectiveness? What skills did you help them develop to excel in their roles?

Example: Providing a standardized communication template that leads to increased stakeholder satisfaction or establishing a TPM onboarding program that enables new team members to ramp up faster.

Your Direct Impact

As a TPM manager, you are an active participant at the organizational level for any strategic decisions and discussions with other peer leaders. How did you use your broader understanding to enable the organization's success?

Example: Identifying and implementing a new tool that would reduce the time to market for product launches or establishing inclusion and diversity initiatives that would lead to meeting hiring goals.

ATTRACT AND RETAIN TOP TALENT

While hiring and recruiting is part of a manager's responsibility, the kind of talent you acquire and retain directly impacts the success of your team and the organization as a whole. TPM leaders who prioritize attracting and retaining top talent create a competitive advantage, foster innovation, and drive exceptional results. In addition to the leadership and manager behaviors mentioned in the previous section, here are three strategies that can enable you to hire and nurture the best talent and reduce attrition:

1. **Invest in your high performers:** Provide opportunities for your star performers by assigning them to innovative programs that will leverage their skills, expertise, and fresh perspectives. Their diverse experience will drive innovation and excellence across the entire team while fostering creativity. Enable their own learning and help them enhance their skills and expand their capabilities. Often managers end up spending more time with team members who may not be performing up to par or need hand-holding. However, your top performers need equal attention and the time you invest in them has a higher impact.

2. **Leverage your top performers:** Make the entire team stronger by having your strongest team members mentor junior or developing TPMs. Ask them to share their knowledge and experience handling tough situations with the community. This way you create a well-rounded team that is ready for the ever-evolving tech landscape. This can also be an opportunity for you to create a leadership bench on your team who can take your role when you decide to move upwards.

3. **Develop a strong TPM brand:** A positive employer brand attracts top talent by showcasing the organization's commitment to employee development. A strong TPM brand establishes trust and confidence regarding career advancement opportunities and a supportive work environment. Engage with the TPM community and network outside your organization to enhance the TPM brand and attract new folks to the team. Attending industry conferences and forums is a great way to get in front of potential candidates and show them the impact of TPMs at the company.

ADDRESS PERFORMANCE ISSUES QUICKLY

In my first six months as a new manager, I had to work through a challenging performance issue on my team. I can tell you from experience that navigating that process was incredibly taxing, especially as a new manager. Since that first time, I have had to do it multiple times but it doesn't get any easier. Performance conversations are tough and there is often fear of conflict and discomfort. This is a sensitive topic where stakes are often high for both parties.

However, it is important that you address performance issues quickly to reduce any negative impact on the rest of your team and the broader organization.

Identify performance challenges: Ideally you already have a pulse on your team's performance through your own observation and conversations with stakeholders. Take a proactive approach, have open conversions, and give timely feedback to your team members, so they can course correct in real time. If issues persist, it's time to investigate the underlying causes.

Diagnose the root cause: When addressing performance problems, don't just treat the symptoms. Dig deep to uncover the root causes of the issues. Is it a lack of skills, unclear expectations, personal challenges, or something else? Pinpointing the cause is essential for effective solutions.

Have a clear performance conversation: Schedule a private meeting with the team member in question and approach the conversation with empathy, compassion, and professionalism. Communicate clearly that this is a performance conversation and you have concerns. Be specific about the problems you've observed, impact on the program or team, and ask for their perspective to gain a full understanding.

Support their development: Once you have aligned with the team member, work together to create a plan for improvement. Set clear goals and expectations, provide resources and support, and establish a timeline for follow-up. Monitor progress closely.

Align and seek guidance if necessary: Maintain open communication and seek guidance when dealing with performance issues. Keeping your superiors and HR in the loop ensures that you have the support and insights needed to address the situation effectively. You may need to keep detailed records of all conversations, feedback, and performance improvement plans. This documentation is essential for tracking progress, justifying decisions, and, if necessary, providing evidence in case of disputes.

Explore alternatives: If, despite your efforts, the team member's performance doesn't improve and it becomes clear they are not a fit for the role, you may need to consider a managed exit. This should be done with sensitivity, following company policies and legal guidelines.

NURTURE TPM COMMUNITY

When I first attended an event organized by an internal TPM community, I felt such a sense of belonging. I felt like all the people were my tribe and they understood me. You see, I was the only TPM on a team of 100+ people. In fact, we were only four TPMs across a 1000-person organization. At times, being a TPM felt lonely because on most days, I was working with engineers, product managers, or other cross-functional partners. Being part of the company-wide communities empowered me in profound ways, like increasing my confidence, because I knew I was not alone dealing with certain situations or motivating me to start a mentorship program for TPMs, and eventually leading me to create a community beyond the company.

As a TPM Leader, your impact extends beyond your immediate team and the larger organization. Build a broader community and actively engage with individuals. Leave a legacy behind and encourage others to do the same.

Knowledge Sharing and Mentorship

Share your expertise and insights with other TPMs within your organization or in the wider professional community. Offer to mentor junior TPMs or those aspiring to become TPMs. Provide guidance, share best practices, and offer support to help them navigate their own career journeys. Actively participate in internal and external forums, conferences, or workshops to present your experiences, lessons learned, and innovative approaches to program management.

Belonging and Collaboration

Take the initiative to build and foster a sense of community among TPM professionals. Initiate or join TPM-focused groups, communities, or networks within your organization or industry. Organize regular meetups, knowledge-sharing sessions, or workshops to facilitate networking, collaboration, and learning opportunities. By bringing together TPMs from diverse backgrounds, you can create a space for shared learning, collaboration, and professional growth. Seek opportunities to collaborate with other TPM Managers or teams across the organization or industry. Share resources, insights, and lessons learned to foster a culture of collaboration and knowledge exchange. By working together, you can tackle common challenges, drive innovation, and elevate the impact of TPM practices across the community or organization.

Thought Leadership and Industry Engagement

Establish yourself as a thought leader in the TPM domain by actively contributing to industry publications, blogs, or podcasts. Share your insights, perspectives, and practical advice on program management, leadership, and industry trends. Engage in discussions and debates on relevant platforms to contribute to the evolution and advancement of TPM practices. By elevating your visibility and credibility as a TPM leader, you can positively impact the perception and understanding of the TPM role within the broader community.

By actively contributing to the TPM community or organization, you become a catalyst for growth, learning, and collaboration. Your willingness to share knowledge and mentor others will have a meaningful impact on TPMs beyond your team. Your thought leadership will help build a TPM brand, enabling TPM functions in other companies to thrive and elevating the entire TPM profession. Finally, your contributions to the broader community will also help you enhance your own professional growth and reputation as a TPM Manager.

* * *

PRO TIP: If you just got into a people management role, hire a leadership or executive coach to help you develop skills and leverage your strengths to make an impact fast. Remember, as you rise up the management ladder, it can get lonely so having a sounding board can really make a difference between understanding people management versus mastering people management.

TPM SPOTLIGHT

VIVIEN TONG, TPM LEADER

Vivien is deeply engaged in supporting the business messaging team at a multinational internet company as a TPM Leader. Her professional journey began as a software engineer and evolved over time through roles like senior engineer and technical lead. Vivien soon realized that her curiosity stretched beyond technical facets, leading her to become a Product Manager. For a decade, she was immersed in the world of search and ranking. In addition, she also explored the startup scene, focusing on platforms, backend infrastructure, and systems. A unique turn of events placed her in charge of billing for advertising programs where she came across the stark disparities in money allocation. Ultimately, she moved into the Product TPM role, which she has embraced and enjoyed.

How was the transition from PM to TPM?

Vivien: The move from Product Manager to TPM was not one I intentionally planned. My attraction to companies with ads publishing channels drove me, and it was at my current company that I encountered the subtle yet crucial differences between these roles. Adapting wasn't without its challenges, as I grappled with delineating boundaries between the responsibilities of a PM and a TPM. Over time, I managed to navigate these complexities, gaining insights into the distinct perspectives each role offers. The key distinction is that Product Managers are accountable for the product's success or failure, while TPMs have influence in shaping that success, concentrating on execution strategy and seamless delivery. TPMs are also not tied to any one team or product, which makes it a flexible role. The TPM's ultimate goal is to deliver the business objective.

What motivated you to stay in TPM?

Vivien: It's true that TPM roles can be demanding due to their inherent lack of complete control over outcomes. There are many pros and cons to staying in the TPM role. What anchors me in the TPM role is the liberty to transcend specific teams or products. This ability to connect different parts of the organization is a unique privilege not afforded to all Product Managers. Moreover, I relish the chance to maintain a strong technical focus, a luxury that isn't always available to PMs.

What surprised you as you navigated the TPM role?

Vivien: Even after ten years in the TPM role, I continue to be surprised by the need to educate others about its nuances. The TPM role remains relatively new, and there's still a considerable lack of awareness regarding its intricacies and contributions.

What do you like about being a TPM?

Vivien: The satisfaction of being an effective TPM comes with recognition and the opportunity to lead substantial, visible programs. This ability to make a tangible impact is a rewarding aspect of the role.

What do you dislike about the TPM role?

Vivien: At times, I find myself wishing for a broader overall understanding of the invaluable contributions TPMs bring to the table. There's room for improvement in recognizing the value they bring using many different skills.

What influenced you to move to a TPM Management role?

Vivien: As my journey progressed, I reached a point where I felt limited by my individual capacity and impact. I aspired to amplify my influence on the business and got the opportunity to mentor and coach fellow TPMs, nurturing their growth and effectiveness. Playing a part in their development and growth was a powerful motivator. The move was not necessarily a great fit from the start, but I learned to embrace the changes that come with TPM management. I know that my impact could be high. Becoming a parent showed me how a people manager is similar in nature. There is a lot of responsibility for others' well-being, which requires a lot of energy and hard work. People management is not just about managing a team's performance and program scale, there is a lot of work behind the scenes that requires time. The management path is not for everyone and it is important to know that there are still many ways to grow as a TPM without choosing people management.

What other differences did you find in being an IC versus Manager?

Vivien: Transitioning from an IC to a manager introduced a fundamental shift. As an IC, my impact was direct and hands-on. In contrast, as a manager, my influence stemmed from guiding and enabling others. Excelling as an IC doesn't necessarily guarantee success in a managerial role. The skills required differ, as managers need to empower and facilitate their team members rather than providing direct solutions. More importantly, transitioning from an IC to a manager is a reversible decision; it's perfectly acceptable to revert to an IC role if needed.

What do you like and dislike about people management?

Vivien: People management involves a series of trade-offs, where the advantages should ideally outweigh the drawbacks. The opportunity to guide and develop a team brings a sense of fulfillment, even though the responsibility can be weighty at times.

What do you think is the best way to leverage TPMs in an organization?

Vivien: In a leadership role, my focus centers on quality over quantity when it comes to TPMs. It's not about rapidly expanding the TPM team or burdening them with numerous projects. Instead, the goal is to align TPMs with their competencies, priorities, and strengths. Quantity isn't the defining factor; what truly matters is the effectiveness of each TPM.

What advice would you give other TPMs or aspiring TPMs?

Vivien: For TPMs and those aspiring to be TPMs, it's crucial to differentiate between the TPM and PM roles. Understand the distinct expectations, goals, and strengths associated with each role. Assess whether your inclination leans toward the product and business side or the technical and execution-oriented aspects. It's important not to view TPM as merely a stepping stone to becoming a PM; recognize and embrace the unique value that the TPM role brings.

PUT IT INTO PRACTICE

✓ Identify your key stakeholders that you need to work with closely to garner support for your long-term goals. Start collaborating with them and understand their motivations.

✓ Make a list of leadership and manager behaviors and identify the ones that you need to practice more often. Identify how you will do it and who you need for support.

✓ Join a TPM community outside your organization and connect with other TPM leaders.

Part 6

SETTING UP TPMS FOR SUCCESS

A NOTE TO LEADERS AND TEAMS WHO WORK WITH TPMS

Coming together is a beginning.
Keeping together is progress.
Working together is success.

– Henry Ford

CHAPTER 26

LEVERAGING TPMS EFFECTIVELY

*We can't solve problems by using the same kind of thinking
we used when we created them.*

– Albert Einstein

Technology today is evolving at a rapid pace and products are getting more complex, which can increase the chances of failures. Managing such complex programs requires someone like a TPM to oversee the entire program to anticipate risks and ensure the product is delivered efficiently at the highest quality. Bringing in TPMs with the right expertise at the start of a critical program can reduce the chances of failure and result in smoother execution. They can help navigate complex requirements, track critical dependencies, and ensure seamless coordination across multiple teams in a matrixed structure. For the best chances of success, organizations must think about how their teams are growing and strategize how they will scale and at what rate. If you are in a growth stage or expecting execution challenges, then it's the right time to hire TPMs.

TPM ALLOCATION

A few years into being a TPM, I joined a team working on the international expansion of a messaging product. The team had 5-7 engineers, one product manager and a couple of other cross-functional partners. Occasionally, the team needed to collaborate with another team like Privacy, Payments, etc. While I enjoyed working on the product and launching consumer centric initiatives, I always felt a sense of unease, like I was not living up to my potential. The work I was expected to do was not challenging enough for me and misaligned with my level. I proactively sought out another program that would feel meaty - something with more complexity and ambiguity. When I took on this huge cross-functional program that required intense problem solving across people and technology, I realized that my previous team did not really need a full-time senior TPM. The engineers could have easily managed new feature launches or maybe a junior TPM would have been sufficient for their needs.

TPMs MANAGE PROGRAMS, NOT TASKS!

This experience made me realize that organizations may not understand if they need a TPM. Later in my career When I was leading TPM organizations, I got many TPM staffing requests from engineering managers. When I asked what the TPM would do, they often indicated that the EM or PM needed support to do their job. Those aren't good reasons because TPMs are paid for delivering complex programs on time and within constraints, not taking on tasks that an EM/PM doesn't want to do. That led me to formalize five program attributes or principles that determine the allocation for a TPM. This will help teams think more proactively about the situations or programs that would be the right fit to bring on a TPM.

A program should have at least have one of these program attributes to request TPM allocation:

1. Complex and large

2. Highly cross-functional

3. Ambiguous

4. Time-sensitive

5. Long-term

Complex and Large Programs Requiring Significant Problem Solving

When facing complex and large-scale programs that involve intricate technical or product challenges, it is essential to consider hiring TPMs to bring a deep understanding of technical intricacies and possess the problem-solving skills necessary to tackle complex issues. Their expertise helps break down complex problems into manageable components, identify potential pitfalls, and drive effective solutions. Hiring TPMs early in such programs ensures the availability of experienced leaders who can navigate the complexities early on rather than fixing problems late in the cycle.

Example: Launching a Smart Assistant like Alexa on a hardware device for the first time.

Highly Cross-Functional Programs with Many Dependencies

Programs that involve numerous dependencies across different teams and functions require effective coordination and collaboration. When teams spend more time managing dependency conversations rather than focusing on their core responsibility, it is probably the right time to bring on an execution expert. TPMs excel in bridging these cross-functional gaps and driving alignment among stakeholders. In doing so, they also help their engineering partners focus on their core strength, which is coding or solving the technical issues. TPMs can communicate effectively across reporting structures, build relationships with partner teams, and negotiate priorities across various teams. This will help organizations ensure streamlined communication, efficient resource allocation, and effective resolution of dependencies, leading to improved program execution.

Example: *Integrating new partners or plugins from third parties like an AI note taker into the existing meeting software like Google Meet or Zoom.*

Ambiguous Programs with a Large Set of Unknown Problems

In scenarios where programs are characterized by ambiguity and a multitude of unknown problems, hiring TPMs becomes crucial. TPMs possess a deep understanding of both the business and technological aspects of the program. Their expertise enables them to anticipate potential challenges, mitigate risks, and guide the team through uncertainty. By leveraging their knowledge and experience, TPMs provide valuable insights, develop contingency plans, and help teams navigate uncharted territories. Bringing in TPMs early in such programs ensures proactive problem-solving, and minimizing the impact of unknowns on program execution.

Example: *Building self-driving cars is highly complex, the technology is still evolving, and not every problem is known yet.*

Time-Sensitive Programs with Little Wiggle Room

When faced with time-sensitive programs that leave minimal room for slippage, hiring TPMs becomes imperative. In fact, TPMs would thrive on having a deadline because they excel in managing tight timelines, tracking progress, and mitigating the risks that may cause any delay. Their strong organizational and time management skills enable them to streamline workflows, identify critical path activities, and proactively address bottlenecks. By involving TPMs in time-sensitive programs, teams can enhance their ability to meet deadlines, mitigate risks of delays, and maintain momentum towards successful program completion.

Example: *Launching highly anticipated hardware devices like smartphones to get ahead of the competition and meeting customer demand.*

Long-Term Initiatives Requiring Multi-Year Planning and Execution

Sometimes, it's easier for teams to manage programs that are three to six months out. Most variables are known and estimates are more accurate. However, for long-term initiatives that span multiple years, there may not be enough clarity upfront. Such programs may require an initial research phase before figuring out milestones or timelines. TPMs can develop a robust planning and execution framework to reduce fuzziness and bring about clarity. They can create comprehensive roadmaps taking both product and technical aspects into account. They can then establish clear milestones and set up effective governance structures. By hiring TPMs early in such initiatives, organizations benefit from their ability to drive long-term strategic planning, ensure alignment with business objectives, and maintain continuity throughout the program's lifecycle.

Example: Building a brand new ecosystem like the Metaverse requires intensive research, understanding of market conditions and technical challenges or possibilities.

* * *

I hope you and your teams will adopt these principles and be proactive in bringing on a TPM for the best chances of success.

PRO TIP: If a team is small enough that it can be fed with two pizzas, (Hern 2018) they can be self-sufficient and do not need a TPM to manage team projects.

CHAPTER 27

STRUCTURING TPM ORGANIZATIONS

Every company has two organizational structures: The formal one is written on the charts; the other is the everyday relationship of the men and women in the organization.

– Harold Geneen

In my first role as a TPM, I was part of a TPM team (reporting to a TPM Manager) that was embedded within the engineering organization. While early in my TPM career, I was fortunate to have a manager who herself had been a TPM for many years. Being in this Program Management Office (PMO) style structure really enabled me to grow my skills and navigate the intricate relationship with engineering. It made my onboarding smooth and I also learned about how one needs to adapt to startups and a bottom-up culture which can be different from traditional top-down hierarchies. My manager was really able to set me up for success, which I will always be thankful for. Contrary to that, when I switched jobs and started reporting to product or engineering leaders in a decentralized format or even as the only TPM, I felt lost. While I knew what I needed to do, I wasn't getting the right support that would help me level up. Over the next few years, I moved in and out of centralized TPM structures and learned to navigate both. Even though my preference leans more towards having a centralized structure, I still believe that there are certain scenarios where a decentralized or hybrid structure is advantageous.

The design of an organizational structure is not a one-size-fits-all solution. There is no perfect org structure. Different companies, teams, and even phases of growth demand varying approaches. It can be necessary to switch between different structures as demanded by a multitude of factors. However, in the ever-evolving landscape of technology companies, the structure of the TPM organization plays a critical role in its success. There are many factors involved in making these decisions. The key lies in regularly reassessing and adapting the structure to meet the needs of the business and the function. The important question to answer here is, "How can TPMs best contribute to the delivery of value to users and the organization?"

One of the critical decisions when designing a TPM organizational structure is determining the factors such as organizational priorities, the need for cross-functional collaboration, and the influence TPMs have on critical decisions.

CENTRALIZED VS. DECENTRALIZED

Centralization and decentralization represent two ends of the organizational spectrum and are commonly referenced for cross-functional organizations like TPM. In a centralized TPM structure, TPMs are grouped under a single umbrella, reporting to a centralized TPM leader. This model offers consistency, standardized hiring practices, and streamlined job functions.

Conversely, in a decentralized TPM structure, TPMs can be embedded within different engineering or product teams, often reporting directly to their respective team leads. This arrangement fosters ownership, enables teams to make autonomous decisions, and ensures that TPMs have a direct seat at the decision-making table.

Both centralized and decentralized TPM structures have their merits and limitations.

	CENTRALIZED	DECENTRALIZED
BENEFITS	• Provides consistency and uniform standards for defining the TPM role and responsibilities. • Helps streamline processes across the organization to reduce duplication of work. • Hiring is more efficient and results in attracting top talent. • Facilitates cross-functional collaboration and TPMs are viewed as a neutral party. • Beneficial for new and junior TPMs who are still learning the trade of being a good TPM.	• TPMs may feel more empowered and tightly connected when embedded within individual teams. • TPMs may feel more ownership, alignment with team goals, and have direct influence on strategic decisions.
LIMITATIONS	• TPMs may find it difficult to get a seat at the decision-making table and influence critical decisions. • TPMs may be viewed as outsiders if the engineering/product teams are not cognizant enough which may impact relationships and collaboration.	• Efforts may be duplicated and there could be inconsistency in performance and hiring standards if the different managers are not communicating with each other. • TPM's career growth may be stagnated due to lack of scope or lack of understanding of what skills will help them grow. • TPMs may feel isolated and may result in slower growth as they won't get opportunities to learn from other TPMs.

Centralized vs. Decentralized TPM Organizations

You can observe that there are more benefits to having a centralized organization, but as I mentioned earlier, decentralized teams can function well if the organization within which it is embedded puts in effort to leverage them in the best possible manner.

If you are contemplating which structure is right for you, here are three recommendations based on my experience as a TPM and TPM leader who has been on both sides.

1. If you are just starting to hire TPMs, embed TPMs with engineering teams.

   Example: *Hiring the first few TPMs at a startup.*

2. Once you have about 5-6 TPMs, start thinking about a centralized structure and having a TPM leader manage those TPMs. You should work closely with the TPM leader to align on the roles and responsibilities.

   Example: *As a startup gets into scale mode and intends to hire more TPMs over the coming months.*

3. If you are a large organization or company with hundreds of TPMs, it may be a good time to think about decentralization. In fact, I recommend a hybrid approach - meaning that small pockets of centralized TPM teams are embedded within large departments. At this point, most standards have been established and you already have strong TPM leaders who can lead these smaller TPM teams. This is something commonly done in big companies like Google or Facebook.

   Example: *Google Ads might have a central TPM team and Google Mobile division might have another. There is no single, monolithic Google TPM organization because it is not feasible to manage such an organization. This will be similar to engineering or product teams where they are aligned with product or business verticals.*

Michael Götz, featured in Part 1, also says that the choice between centralized and decentralized TPM teams depends on the size and growth stage of the organization, as well as the number of VP+ level stakeholders involved. According to him, decentralized teams may lead to a silo effect, while centralized teams can foster cross-team collaboration and consistency in TPM processes and practices. We both agree that it's essential to strike a balance between centralization and decentralization based on the organization's needs and goals. A larger company with many stakeholders may benefit from centralized TPM teams to streamline communication and alignment. Smaller organizations might find decentralized teams more suitable to maintain flexibility and adaptability. Ultimately, the decision should align with the company's culture, structure, and strategic objectives. Both models have their pros and cons, and the best approach is to choose one that suits the organization's unique requirements at a certain growth stage.

SETTING THE TEAM UP FOR SUCCESS

Irrespective of which structure you choose, spend some time and effort aligning on how best to set up the TPM organization for success. For a centralized TPM structure to succeed, it requires strong alignment with business needs, efficient hiring processes, well-defined career paths, and consistent communication across teams. The centralized approach is more conducive to building a highly specialized and powerful TPM function that spans across different domains.

In contrast, a decentralized TPM structure necessitates robust mentorship and community-building efforts. TPM leaders must facilitate open lines of communication between embedded TPMs to avoid isolation and promote knowledge sharing. The success of a decentralized structure hinges on engineering and product leaders understanding the value TPMs bring to their teams and the organization.

* * *

Designing an effective TPM organizational structure is an ongoing process that demands a deep understanding of the business, the function's role, and the team's evolving needs. A balanced approach that considers both business priorities and the unique contributions of TPMs is key to creating a successful and impactful TPM organizational structure.

PRO TIP: The decision between centralized and decentralized structures is not absolute; it requires continuous evaluation and adaptation. So don't be afraid to switch from one to the other if the business needs have changed.

CHAPTER 28

PROVIDING TPMS VALUABLE FEEDBACK

We all need people who will give us feedback.
That's how we improve.

– Bill Gates

A few years ago, I led the launch of an AI-driven product on a hardware device. This was a highly complex program that went through its share of challenges. However, we were able to launch it successfully in the desired timeframe. I got the opportunity to really stretch myself and go above and beyond. For the performance review cycle that year, I asked one of the product managers I worked closely with for feedback. The reviewers have the option to share their feedback with you. While I got a great review, I was highly disappointed in the review by this partner because it did not call out my impact or the most challenging issues I was able to resolve. Instead it mentioned how I got some "phone booths" setup so QA was able to do their testing. Now granted, that was part of the job and unblocked testing, but there were bigger and more impactful contributions I made that were overlooked. That review made me realize that while everyone knows "feedback is a gift," most people find it difficult to give good and valuable feedback. Many cross-functional partners may not understand enough to give feedback that will help TPMs improve or give them recognition of the right things.

Providing effective feedback for TPMs requires a deep understanding of their role, skills, and contributions. In many cases, teams and individuals may not fully grasp how TPMs are evaluated or what to expect from them based on their seniority. This lack of clarity can result in feedback that is not useful for the TPM's growth.

Here are a few guidelines on delivering actionable and valuable feedback to TPMs. Doing so will help the TPMs thrive in their roles and make meaningful contributions to program success.

UNDERSTAND THE CORE TPM SKILLS

To provide valuable feedback to TPMs, it is crucial to have a solid understanding of the core skills that define their role. As outlined in Part 2 of the book, these skills include technical expertise, program management, leadership, and communication. Familiarizing oneself with these evaluation axes sets the foundation for providing targeted and constructive feedback that aligns with the expectations and responsibilities of TPMs.

Key Questions to Draft Actionable and Valuable Feedback

To facilitate the process of delivering feedback, consider asking yourself the following key questions:

What key results did the TPM achieve?

Assess the specific outcomes and achievements the TPM has contributed to. Look for tangible results that demonstrate their effectiveness in driving successful program execution.

- **What business outcomes did the TPM drive and influence?** Evaluate the impact of the TPM's work on overall business objectives. Identify how their contributions have influenced important metrics, revenue growth, customer satisfaction, or other key performance indicators.

- **If no TPM was present, what would have been their impact on the business?** Consider the hypothetical scenario where the TPM was absent from the program. Reflect on the potential consequences and challenges that would have arisen without their presence, highlighting the value they bring to the organization.

- **What contributions did the TPM make towards organizational efficiencies?** Examine how the TPM has improved processes, streamlined workflows, or enhanced efficiency within the organization. Recognize their efforts in driving operational excellence and optimizing resources.

How did the TPM achieve these results?

What core skills did the TPM demonstrate and what strategies or methodologies did they employ to achieve the results described? Check out the following examples:

- Take ownership, drive accountability towards agreed-upon deliverables, and set clear expectations and deadlines.

- Identify and resolve problems, dig deep to analyze root causes, and find effective solutions quickly.

- Balance their capacity, understand the big picture while executing day to day activities.

- Align key cross-functional partners to build broad consensus on dependencies and roadmap delivery.

- Bring in innovative ideas to technical or product discussions and enable the team to think outside the box.

- Communicate progress in a way that provides clarity and confidence in their actions.

* * *

Your feedback will enable TPMs to further develop their skills, enhance their effectiveness, and excel in their roles, ultimately contributing to the success of the programs on your team.

PRO TIP: If you have a TPM Manager for a TPM on your team, work with them to understand what kind of feedback would best help them and their team. Or check with TPM leaders from outside your team or organization.

CHAPTER 29

OPERATING AS ONE TEAM

We may have all come on different ships,
But we're in the same boat now.

– Martin Luther King, Jr.

Tpms work very closely with engineering and product teams on a day-to-day basis. TPMs are also working in matrixed structures across multiple teams, so they may not "belong" to any one team. It is easy to forget about the TPM on important occasions like team-building activities or All Hands shout outs. TPMs miss opportunities for building connections with the people they most closely work with and can often feel left out or undervalued. Giving TPMs the same recognition and visibility can result in more seamless collaboration and team engagement. Integrating TPMs as core members of the team can increase team effectiveness and engagement.

TPM ONBOARDING

If you are ready to bring on one of your first TPMs aligned with the TPM allocation principles we discussed in Chapter 26, here's how you can ensure that they are successful and enable you to meet broader business objectives.

Identify priorities: When bringing on a new TPM, have a plan in place before they join your team and organization. Think about your top three priorities that need TPM support. Talk to your team members and other stakeholders and get them prepared for the TPM's arrival. Make sure everyone on the team understands the TPM's role.

Work together: Onboard the TPM and support their integration into the team during the initial days. The TPM will likely have lots of questions and having someone to partner with will help them understand the team and organization better and ramp up faster. This will also make it easier for them to find the right solutions that work for you.

Empower them: Give the TPM ownership a meaty problem. Point them in the right direction but empower them to come up with ideas and solutions. Brainstorm and discuss those ideas together and even challenge them to think outside the box. Hold them accountable to the expectations that you set for them. Build a trusting relationship and the TPM will be your biggest advocate.

Involve them: TPMs are often analytical and need to understand the rationale behind what is being asked of them. Make sure that TPMs are in all relevant product and technical conversations like requirements discussion, design reviews, user feedback, etc. They may not be writing requirements or code, but their understanding of the big picture and ability to connect the dots is highly useful in all conversations.

Provide clarity on ownership: We all know the TPM role straddles multiple disciplines, especially product and engineering. In smaller teams, an engineering manager or product manager will often play the TPM role. However, as teams grow and a TPM is brought on to handle cross-functional complexity, the lines of ownership of certain areas can get fuzzy. Every team is different and every individual has different dynamics. Therefore, it is important to sit down and have an open discussion on ownership and who will do what.

Provide timely and direct feedback: A TPM will grow and make a bigger impact with continuous feedback. Provide them with feedback regularly, so they can quickly incorporate any changes in style or action to produce results that work for the team. Make your feedback direct and honest. This can be difficult but circumventing the actual issue will not help either party.

Following these guidelines will help you leverage them effectively for the overall success of the organization.

TPMS AS STRATEGIC PARTNERS

One of my biggest pet peeves around TPMs is considering them as a support function. Every function - engineering, product, design, sales, marketing and so on - is equally important and has its own unique strengths that enable the delivery of business goals. If you are renovating your house, you wouldn't want the electrician to do the plumbing, would you?

Consider TPMs as strategic partners from the outset, involving them in key decision-making and strategy discussions, and recognizing their unique contributions to program success. By doing so, you will unlock their full potential to drive innovation and deliver successful products. Leverage their distinct skill set and unique perspective that combines technical expertise, program management skills, and leadership abilities. Their role extends beyond traditional project management, as they are responsible for driving the execution of complex initiatives, coordinating cross-functional teams, and aligning program objectives with organizational goals.

Keep these four tips in mind to build successful partnerships with TPMs:

Including TPMs in Strategy and Planning

Planning is a critical phase in product development as it sets the direction and vision for the entire program. To ensure comprehensive and well-informed strategies, it is essential to include TPMs in the formulation process. Their deep understanding of technical complexities, market dynamics, and program execution challenges enables them to provide valuable input on feasibility, risks, and potential opportunities. It will also lead to smoother execution as the TPM understands the overall picture and has developed more robust and realistic plans that align with organizational goals.

Integration in Key Decisions and Discussions

TPMs have keen technical acumen, program management expertise, and cross-functional understanding, making them valuable contributors to strategic discussions. Involving TPMs in key decisions and discussions can ensure a holistic and comprehensive approach to program execution, providing insights that address potential challenges, optimizing resource allocation, and aligning program objectives with business goals.

Collaborative Approach to Problem-Solving

TPMs excel in navigating complex challenges and bridging gaps between different functional teams. By engaging them early in problem-solving processes, organizations can tap their ability to identify potential roadblocks, understand dependencies, and propose creative solutions. By operating as one team - with TPMs as equal strategic partners - organizations can foster a culture of collaboration, where expertise from all disciplines is leveraged to drive innovation and overcome obstacles.

Driving Alignment and Accountability

By considering TPMs as strategic partners, organizations foster a sense of ownership and accountability across the entire team. When TPMs are actively involved in decision-making, their understanding of program goals, technical constraints, and cross-functional dependencies helps them align teams and stakeholders. This alignment promotes a shared sense of responsibility and empowers all team members to contribute to the success of the program.

* * *

It is essential to view TPMs as part of the team and give them a platform to share their valuable insights. Don't just view them as a support function, but think of them as integral to shaping the success of your organization's product development endeavors.

> **PRO TIP:** Anytime you partner with a new TPM on the team, talk to them about each of your roles and responsibilities. Discuss complementary skills and an engagement structure that will support the partnership. While RACI charts can help, having a 1:1 conversation can establish trust and give you a deeper understanding of how the TPM operates. *Note: I recommend the TPMs do the same whenever they start on a new team or program.*

TPM SPOTLIGHT

AMANDA DONOHUE, TPM LEADER

Amanda is a technology leader at a multinational financial software and services company. She has had an incredibly rewarding experience being a part of the company's growth that led her to explore different roles including leading TPgM function, product operations, collaboration engineering and, most recently, people programs and development. Prior to that, she was at a big tech company for a significant part of her career where she once again took on various roles, starting from the ground up as an entry-level IC handling tasks like resetting passwords. Over the years, through pivotal moments and very supportive management, Amanda moved from operations to the world of technical program management. She became a respected TPM leader overseeing highly technical teams within infrastructure organizations. Amanda's career transitions have been driven by the desire to diversify her skill set and contribute to building something new and impactful.

Can you share more about your move to one of the most valuable startups after spending many years at a big tech company?

Amanda: This new opportunity provided me with a unique chance to start something new and build a centralized TPM function from the ground up. This experience allowed me to develop totally different skill sets and operate as a leader in a high growth environment. It has also been an interesting journey because the role was not fully defined, which allowed me to be more creative and define a path forward.

Your journey seems to have taken quite a few turns. Can you shed light on how you transitioned from non-technical roles to becoming a TPM and eventually leading teams?

Amanda: It was an interesting evolution. I started in an entry-level role, quite far from where I am now. I was exposed to technical work in mobile engineering in those early years, which I enjoyed. I continued to learn by observing and digging deep to figure out where I can add the most value. My initial foray into TPM was accidental, and at the time there wasn't a clear definition of the role. I found my niche in release engineering while at this big company by really getting into the weeds and deeply understanding what it takes to launch high impact technical programs. I leveraged that experience to grow into a technical leadership role. My journey underscores that being technical is more than just coding. I can be at the intersection of technical programs where you need deep program management skills, understanding how to work with engineering and leverage leadership skills to move forward.

What is the biggest factor that enabled your leadership journey?

Amanda: In the early days, I struggled with confidence because not having a technical background in the traditional sense felt like a chip on my shoulder. What truly helped me was having mentors and sponsors who believed in me. I had managers and leaders who invested time in my development, which helped me build confidence. These relationships didn't start with formal mentorship discussions, but rather, they developed organically through collaborative work and shared goals. Having leaders who gave me opportunities, guidance, and constructive feedback greatly shaped my career trajectory and the opportunity to discover what leadership meant to me.

What were some of the challenges you faced as you took on leadership roles?

Amanda: For many years, I was working in a mature environment where the TPM job ladder and level expectations were established. In my new role as the head of TPgM, I realized that as I was building things from scratch, all decisions came down to me. This was a scary moment - how do I know if I am making the right decisions? It also takes a while to see the results of your actions, unlike shipping an update. I found that leaning into collaboration and building strong relationships with peers and colleagues helped. Also, recognizing that learning from mistakes is part of the process.

What advice would you give rising TPM Leaders?

Amanda: My advice would be twofold. First, find your niche within the TPM role. Whether you're more inclined towards program management, technical leadership, or another area, lean into what you're passionate about and where you can make a unique impact. What skills do you bring to the table and how can you continue to deepen those skills? Second, as you grow, you might need to step back from certain technical tasks but don't lose touch with those skills. At many of the smaller companies, you are probably going to need to be hands-on, which means you may need to run programs. So understand how you can balance that while building and leading a team. This will help you build a deeper understanding of the system and empathy for your team.

💬💭 IN CONVERSATION WITH TPM ALLIES

SIWEI SHEN, ENGINEERING LEADER

Siwei is an engineering leader at an established Crypto company, overseeing the blockchain enablement and staking organizations. Siwei has previously worked at multiple big tech companies and transitioned into an Engineering Manager role about a decade ago. Since then he has been working with TPMs on various programs and initiatives. He has seen how a well-structured TPM function can operate and has also seen that even the most established companies can take time to establish a good structure.

Tell us more about your experience working with TPMs?

Siwei: I believe successful partnerships with TPMs begin with leadership involvement. Organizational structure plays a significant role too, with top leaders needing to comprehend the unique value of TPMs. I've enjoyed working with TPMs, particularly when they are part of a mature standalone organization, each with clearly defined roles. They provide strategic collaboration, bridging different teams that might lack a direct one-to-one mapping, thus reducing dependencies, especially at the senior level.

What is your most memorable experience with a TPM?

Siwei: The remarkable thing about TPMs is their fluidity in roles, adapting to various scenarios. Their technical knowhow, flexibility, and ability to extend processes beyond engineering are impressive. For instance, during a compliance initiative, a TPM extended an engineering-only process to a company-wide incident management process, including the expected involvement from non-technical roles as well. This scenario-specific innovation demonstrated the value of a skilled TPM.

What do you think are the superpowers of a TPM?

Siwei: TPMs possess several superpowers. They're proactive, defining their roles and contributions independently. Their impact transcends technical skills, making them valuable across roles like engineering, product management, operations, and marketing. Importantly, they bring an unbiased perspective, especially when not directly reporting to engineering or product teams.

When working with TPMs, what signals indicate partnership issues?

Siwei: Some signs of a partnership not working optimally include excessive process orientation, overly emphasizing technical aspects, or trying to control everything. These issues can be addressed through coaching, adapting process adherence to company culture, and offering the right organizational support. Successful TPM collaboration requires balancing structure with flexibility.

Based on your experience, when is the optimal time to bring in a TPM?

Siwei: The ideal time to involve a TPM is when a company is mature enough to constantly have initiatives across the team and/or functional boundaries. At this point, domain context outweighs rigid team boundaries. TPMs can be affiliated with initiatives or teams without permanent attachment, allowing for dynamic involvement.

What advice would you give organizations hiring TPMs?

Siwei: One key lesson is that organizations should be cautious when attempting to fit the TPM role within a product group or business unit model. Having TPMs report directly to the team they support optimizes for the short-term efficiencies but can hinder the long-term career development of TPMs.

What advice would you give to aspiring and existing TPMs?

Siwei: Understanding your unique value add is crucial. Technical contributions can vary widely, and even non-engineering backgrounds can offer domain-specific technical insight. Recognize the distinctiveness of the TPM role within a tech company. Set expectations for your role within the EM and PM partnership, emphasizing complementarity. Prioritize tasks based on your strengths and the initiative's needs. Find the right balance between breadth and depth of skills to make the right level of impact.

LISA HUANG, PRODUCT MANAGEMENT LEADER

Lisa is an accomplished product leader with a solid track record in leading teams across both software and hardware with a primary focus on AI/ML domains She has worked at startups and most of the top five big tech companies, so she has seen a wide range of approaches to product building and team setups.

Could you share your experiences working with Technical Program Managers (TPMs)?

Lisa: Having a TPM on board has been immensely valuable for me. Personally, I'm not the most organized individual, so having a TPM to handle organizational aspects has been a game-changer. Beyond that, TPMs play a crucial role in freeing up cross-functional capacity within teams. The best TPMs excel at maintaining order, anticipating challenges, arriving ahead of potential issues, and offering insights into what might slip and how to mitigate such situations. They are adept at flagging risks and managing escalations, providing a clear and comprehensive visibility into the program's progression.

Can you share a memorable experience involving a TPM?

Lisa: One of the most memorable instances was during the MVP launch of a hardware device. The entire process was chaotic–product definitions were in flux–and processes were not well-established. With the assistance of our TPM, we managed to lay out a clear plan detailing what was to be delivered. This alignment and effective communication was crucial in steering the program forward. In another scenario, when we were working on another consumer tech product launch, things were quite challenging. However, having strong TPMs in place made the situation less chaotic. Their ability to synthesize complex information and provide clear visibility proved crucial in managing the program and keeping it on track.

When working with TPMs, what signals indicate that the partnership may not be working well?

Lisa: When partnerships turn problematic, it's often due to overly prescriptive planning. If things become too rigid and predefined, they tend not to work well. It's important to ensure that there's mutual buy-in from the TPMs and other functions. All parties need to understand the reasoning behind the decisions being made. Additionally, a soft influence approach often yields better results than a purely authoritative stance.

Based on your experience, when do you believe is the optimal time to bring in a TPM?

Lisa: The decision to bring on a TPM hinges on having sufficient scope in both the "what" and "how" of a program. If either aspect is lacking, it might not be the right time. Moreover, clarity regarding ownership is really important. It's also crucial to adjust the cultural approach based on the defined role, ensuring that there isn't too much overlap with other roles. Particularly with complex products, as teams grow in size and collaboration becomes more intricate across cross-functional teams, a well-functioning TPM organization can make a significant difference. Striking the right balance is key. Too much process can become a burden, leading to dissatisfaction, but having the right help in place is essential, especially where people require assistance.

PUT IT INTO PRACTICE

✓ If you have a TPM on your team, identify if their program has at least one of the five attributes for allocation. If not, identify another program that is a better fit for bringing them on.

✓ Connect with at least one TPM leader in your organization or outside to learn more about the role and function.

✓ Identify situations or meetings where you can integrate a TPM like technical design reviews or product requirements reviews.

EPILOGUE

EMBRACE *THE ART OF STRATEGIC EXECUTION*

As I reflect upon the journey that we embarked on together through the pages of *The Art of Strategic Execution*, I am filled with a profound sense of gratitude. This book has been my humble endeavor to share insights, experiences, and strategies that can empower aspiring and seasoned TPMs to build successful careers.

Throughout the exploration of the TPM landscape, we explored various aspects of this multifaceted role. We unraveled the importance of cultivating leadership and people skills, understanding the nuances of effective communication, navigating complexity, and fostering cross-functional collaboration. We explored the intangible qualities that distinguish exceptional TPMs—qualities such as adaptability, resilience, and the ability to influence and inspire others.

But the journey doesn't end here. As TPMs, you are lifelong learners, constantly adapting to the ever-changing landscape of technology and program management. The insights shared in this book serve as a solid foundation upon which you can continue to build and refine your skills. Embrace the opportunities to grow, seek new challenges, and push the boundaries of your capabilities.

Remember, success in Technical Program Management is not solely defined by your individual achievements but by the impact you have on your teams, organizations, and the broader community. As TPMs, you have the privilege and responsibility to drive positive change, to lead with empathy, and to foster an environment that encourages innovation and collaboration.

In closing, I want to express my deepest gratitude for embarking on this journey with me. Your commitment to your personal and professional growth is an inspiration. As you forge ahead, may you find fulfillment in the challenges you overcome, the connections you foster, and the impact you make. Remember, your career in Technical Program Management is not merely a destination; it is an ongoing voyage of discovery and growth. I hope this book has illuminated your path and inspired you to embrace *The Art of Strategic Execution*.

Thank you, and may your path be filled with endless possibilities and remarkable achievements.

Wishing you continued growth and fulfillment,

Priyanka Shinde

TPM Leader, Executive Leadership Coach, Keynote Speaker

RESOURCES

Throughout the book, I have mentioned a few complimentary resources that might be helpful to you. Check them out.

The Art of Strategic Execution Toolkit: If you liked the templates referenced in this book, you can now leverage them to streamline your execution. These templates were created over the course of my TPM career and I am excited to share them with you. The special Art of Strategic Execution Toolkit is available exclusively for you. Get it at https://www.artofstrategicexecution.com.

TPM Academy: If you are interested in nailing your TPM interviews, advancing your career, growing as a leader, or even mastering TPM management, then TPM Academy is your gateway to further your learning and become part of the broader community. You will find multiple TPM resources, so you can choose the one that best works for you at this stage of your career. Check out https://www.thepriyankashinde.com/tpm-academy.

TPM Blog: If you are interested in exploring more topics and content related to being a rockstar TPM, then check out my blog. I post about various TPM and leadership topics. Learn more at https://www.thepriyankashinde.com/tpm-blog.

E-newsletter: If you'd like to stay in touch, please join the thousands of readers I write to regularly by signing up for my e-newsletter at https://thepriyankashinde.com/subscribe. I try to make the newsletter personal and useful, so I hope you'll join!

thepriyankashinde.com: Continue your journey with me beyond the book. Join my community of like-minded professionals who are strategic about unleashing their success. If you are ready to embark on a transformative journey, check out my coaching services and how they can support your professional growth.

Visit my website at https://thepriyankashinde.com.

ACKNOWLEDGEMENT

Writing a book is a journey that takes more than just the author's efforts. It is a collaborative effort, and I am profoundly grateful to the individuals and groups who have supported me throughout this endeavor.

First and foremost, I want to express my deepest gratitude to my family - for their unwavering support, understanding, and belief in me. Your encouragement kept me going even when the path seemed challenging.

Sincere thanks to the amazing industry leaders who took the time to review advance copies of my book. Your insightful reviews and endorsements have been truly encouraging.

A heartfelt thank you to Amanda Donahue, Lisa Huang, Siwei Shen, Ben Gauthier, Malvika Sinha, Won Choe, Michael Götz, Vivien Tong, and Gil Segev—who were kind enough to share their experiences and views with my audience. Without their perspective, this book would be incomplete.

Special shout-out to my beta readers who took the time to review the manuscript and provided valuable feedback. Your constructive criticism and honest opinions played a key role in polishing up my ideas and making this book even better.

I am indebted to the team that brought this book to life – from the brilliant editor/s Sandra Huffman and Kathleen Ralls who meticulously combed through the manuscript to the talented designer Dino Marino, who turned my vision into a beautiful book cover. Your professionalism and expertise have elevated this book beyond my expectations.

My gratitude extends to all the people I have encountered in my career journey and who have taught me in unexpected ways. Your partnership and collaboration will always be cherished.

Lastly, to my readers – without you, this journey would not be complete. Your enthusiasm and support inspire me to continue sharing stories.

Thank you, each and every one of you, for being a part of this incredible journey.

With sincere appreciation,

Priyanka Shinde

NOTES

Chapter 1

- Lopp, Michael. 2013. "Entropy Crushers – Rands in Repose." Rands in Repose. https://randsinrepose.com/archives/entropy-crushers/.
- Carayannis, Elias G., Young H. Kwak, and Frank T. Anbari, eds. 2005. "Brief history of project management." In The Story of Managing Projects: An Interdisciplinary Approach, 1-9. The George Washington University: Bloomsbury Academic.

Chapter 2

- Rehkopf, Max. n.d. "What is a scrum master? [+ Responsibilities]." Atlassian. Accessed November 1, 2023. https://www.atlassian.com/agile/scrum/scrum-master.
- "What Is a Scrum Master (and How Do I Become One)?" 2023. Coursera. https://www.coursera.org/articles/what-is-a-scrum-master.

Chapter 8

- Mongan, John, Eric Giguère, and Noah Kindler. 2013. Programming Interviews Exposed: Secrets to Landing Your Next Job. N.p.: Wiley.
- Kleppmann, Martin. 2017. Designing Data-intensive Applications: The Big Ideas Behind Reliable, Scalable, and Maintainable Systems. N.p.: O'Reilly Media.

- Ng, Andrew. n.d. "Supervised Learning in Machine Learning: Regression and Classification (DeepLearning.AI)." Coursera. Accessed November 6, 2023. https://www.coursera.org/learn/machine-learning.
- "Systems design." n.d. Wikipedia. Accessed November 1, 2023. https://en.wikipedia.org/wiki/Systems_design.
- Chakraborty, Ashis. 2020. "System Design 101. Step by step guide on designing a… | by Ashis Chakraborty." Towards Data Science. https://towardsdatascience.com/system-design-101-b8f15162ef7c.
- Grokking Modern System Design Interview for Engineers & Managers." Educative, https://www.educative.io/courses/grokking-the-system-design-interview?aff=x2aM.
- Haq, Fahim U. 2017. "Top 10 System Design Interview Questions for Software Engineers." HackerNoon. https://hackernoon.com/top-10-system-design-interview-questions-for-software-engineers-8561290f0444.

Chapter 13

- Watkins, Michael D. 2013. The First 90 Days: Proven Strategies for Getting Up to Speed Faster and Smarter. N.p.: Harvard Business Review Press.
- Liu, Deb. 2021. "Make the First 90 Days Count - by Deb Liu - Perspectives." Deb Liu | Substack. https://debliu.substack.com/p/make-the-first-90-days-count.

Chapter 15

- "New Year's Resolution Statistics (2023 Updated)." n.d. Discover Happy Habits. Accessed November 1, 2023. https://discoverhappyhabits.com/new-years-resolution-statistics/.

Chapter 16

- Sahadi, Jeanne. 2018. "Sheryl Sandberg: Inside the mind of Facebook's COO." CNN. https://www.cnn.com/2018/10/03/success/sheryl-sandberg-profile/index.html.

- MacKay, Jory. 2019. "The Myth of Multitasking: The ultimate guide to getting more done by doing less." Rescue Time : Blog. https://rescuetime.wpengine.com/multitasking/#multitasking.
- Murty, Rohan N., Sandeep Dadlani, and Rajath B. Das. 2022. "How Much Time and Energy Do We Waste Toggling Between Applications?" Harvard Business Review. https://hbr.org/2022/08/how-much-time-and-energy-do-we-waste-toggling-between-applications.
- Asana, Team. 2022. "Be Productive at Home: 11 Tips to Promote Efficiency [2023]." Asana. https://asana.com/resources/eisenhower-matrix.
- Buckingham, Marcus. 2015. Standout 2.0: Assess Your Strengths, Find Your Edge, Win at Work. N.p.: Harvard Business Review Press.

Chapter 22

- "ICF, the Gold Standard in Coaching | Read About ICF." n.d. International Coaching Federation. Accessed November 1, 2023. https://coachingfederation.org/about.
- Grove, Angela. 2019. "The difference between a coach, mentor and sponsor." Angela Lovegrove. https://angelalovegrove.com/2019/10/07/the-difference-between-a-coach-mentor-and-sponsor/.
- Osmani, Addy. 2022. "AddyOsmani.com - A coach, a mentor and a sponsor." Addy Osmani. https://addyosmani.com/blog/mentor-sponsor-coach/.

Chapter 25

- Nolan, Tom. 2017. "The No. 1 Employee Benefit That No One's Talking About." Gallup.com. https://www.gallup.com/workplace/232955/no-employee-benefit-no-one-talking.aspx.
- The Center for Leadership Studies, founded by Dr. Paul Hersey. n.d. "Understanding the Situational Leadership®

Model." The Center For Leadership Studies. Accessed November 1, 2023. https://situational.com/blog/what-is-situational-leadership-understanding-this-leadership-model/.

- Bungay Stanier, Michael. 2016. The Coaching Habit: Say Less, Ask More & Change the Way You Lead Forever. N.p.: Page Two Books, Incorporated.

- Goler, Lori, Janelle Gale, and Adam Grant. 2018. "Why People Really Quit Their Jobs." Harvard Business Review. https://hbr.org/2018/01/why-people-really-quit-their-jobs.

Chapter 26

- Hern, Alex. 2018. "The two-pizza rule and the secret of Amazon's success." The Guardian. https://www.theguardian.com/technology/2018/apr/24/the-two-pizza-rule-and-the-secret-of-amazons-success.

ABOUT THE AUTHOR

Priyanka Shinde is a seasoned Silicon Valley tech leader with over 20 years' experience in engineering, product management, and program management. She has previously worked at renowned companies such as Meta and startups like Cruise. Throughout her career, Priyanka has consistently delivered strategic business outcomes, leading highly technical teams to successfully launch cutting-edge products in the field of Autonomous Vehicles, Artificial Intelligence (AI), Machine Learning (ML), Augmented and Virtual Reality (AR/VR), and Advertising Tech. Priyanka received her MBA from North Carolina State University and Masters in Computer Science from Arizona State University. Her passion for maximizing TPM excellence and fostering a culture of connection, authenticity, and empathy in organizations led her to become a certified executive coach for tech leaders who want to accelerate on their ambitious goals.

As a thought leader in the industry, Priyanka is highly sought-after as a speaker and trainer. She has graced industry conferences and events, sharing her expertise on a wide range of topics, including achieving excellence, leadership development, and emerging technologies. Her insights and experiences inspire others to strive for greatness and embrace continuous growth in the rapidly evolving tech landscape. Priyanka's expertise extends beyond her roles in engineering and leadership. She specializes in optimizing product, process, and people strategies, making her an invaluable advisor to startups in the tech industry. By closely collaborating with founders and CEOs, she helps organizations enhance their efficiency, effectiveness, and productivity. Priyanka's holistic approach ensures that these improvements are aligned with the core mission, values, and culture of the organizations she advises. Her belief in the power of human connections and building strong relationships has been instrumental in driving success and creating an environment conducive to collaboration and innovation.

Priyanka resides in the vibrant San Francisco Bay Area with her loving family, finding inspiration in the breathtaking landscapes and innovative atmosphere that surround her. She is a certified group fitness instructor empowering others to achieve their fitness goals and lead a balanced lifestyle. Priyanka loves hiking, running and immersing herself in the beauty of nature. Priyanka channels her creativity through baking and painting.